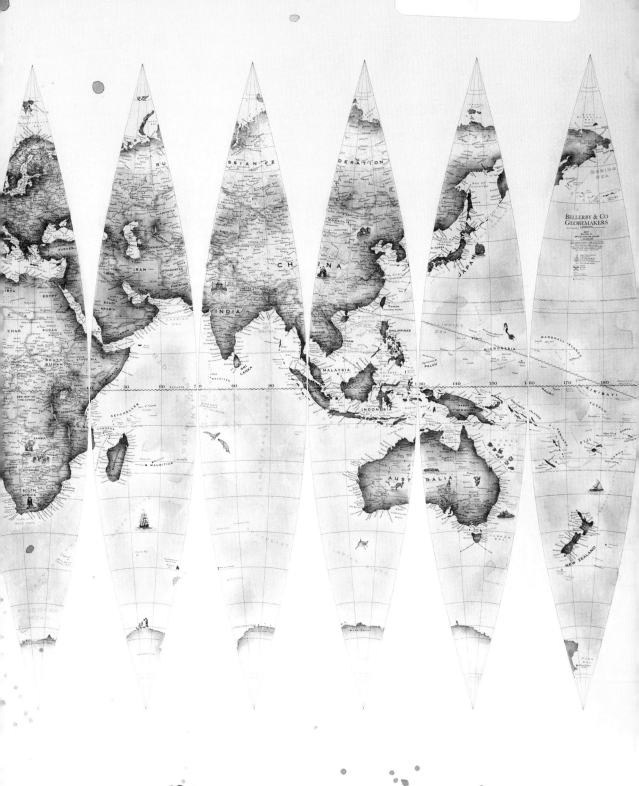

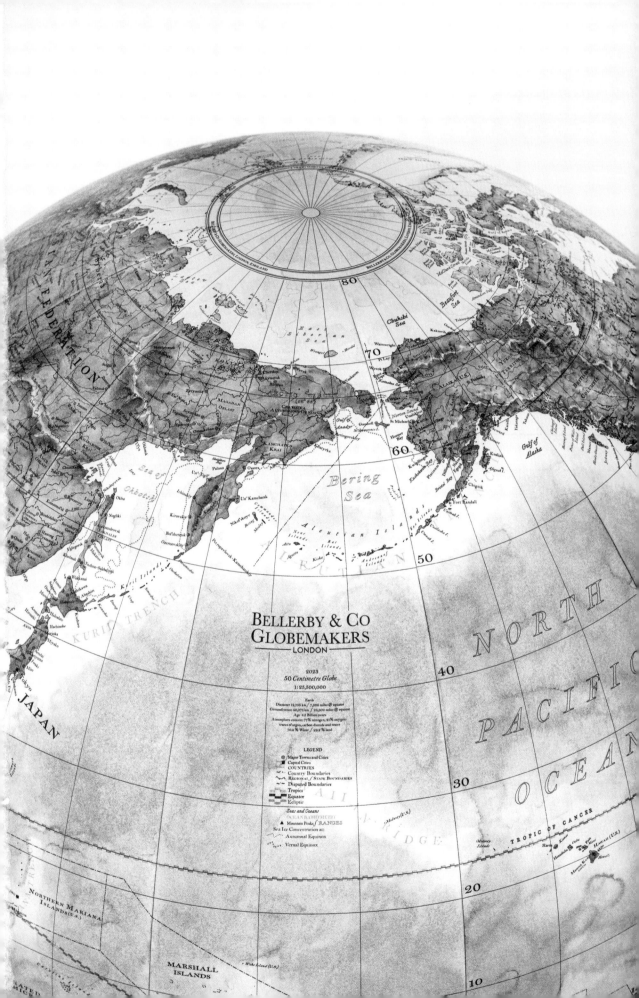

For Mary Anne, Jade's grandma, an amazing and inspiring woman who passed away in early 2021 at the age of ninety-six. Six years earlier, at her ninetieth birthday celebration, her sons had had to persuade her that skydiving was an inappropriate 'experience' for someone of her age, and so she settled for piloting a biplane like the ones she had flown just after the Second World War during her US Air Force service. Her stories about her many adventures were amazing, even when on one occasion at the Bel Air Hotel I had to stop her mid-flow as her hair had caught on fire from a nearby heat lamp. She just laughed! The last time I saw her at her wonderful house with views across the LA canyons, she leaned in, took my hands in hers and simply asked, 'Are you happy?', reminding me that this is the most important thing.

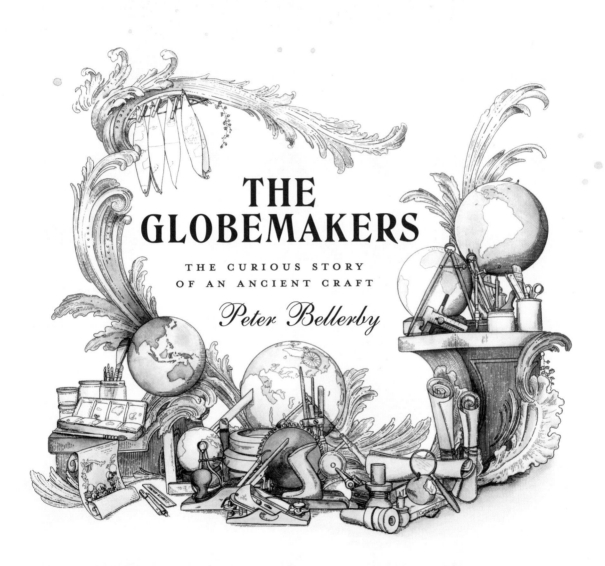

THE GLOBEMAKERS

THE CURIOUS STORY
OF AN ANCIENT CRAFT

Peter Bellerby

BLOOMSBURY PUBLISHING
LONDON · OXFORD · NEW YORK · NEW DELHI · SYDNEY

CONT

ENTS

PREFACE

The accuracy, function, look and feel of globes may have changed over time, but our fascination with them has not. This is the story of how my simple idea to pass some time in relearning an ancient craft evolved into a company, Bellerby & Co. Limited, and the lucky coincidences that shaped it. Along the way I have included some relevant information on historical globes and globemaking, interesting facts and information about our planet, beautiful illustrations drawn by our in-house artists, accompanied by a photographic journey around our studio.

GLOSSARY

Armillary Sphere An astronomical device (also called a spherical astrolabe) representing the celestial bodies, with the Earth or the sun at its centre, and which has a series of concentric rings that correspond to lines of celestial longitude and latitude. Other astronomical features are usually included. A Ptolemaic armillary sphere has the Earth at its centre, and this helps to determine the position of celestial bodies and demonstrate the motion of the planets around it. An armillary sphere can also be constructed with the sun at the centre, when it becomes a Copernican.

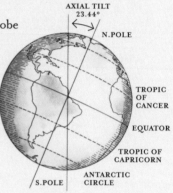

Axis The line around which the Earth rotates. On a globe the axis will often be represented by a metal rod affixed at either end to a meridian, allowing the globe to be spun and thus mimic the Earth's movement. The axis of the Earth is tilted at approximately 23.4 degrees, an angle which oscillates over 41,000 years, and has the North and South Poles at either end. The Earth currently takes twenty-four hours for a full revolution around its axis (around 4.5 billion years ago it was under five hours).

Calotte A skullcap worn by Roman Catholic priests, but in globemaking they are the final round pieces of map you apply to the globe on the North and South Pole. They neatly finish off the globe, while also protecting and securing the tips of the gores in place.

Cartouche This is the name given to the maker's logo, usually placed in the northern Pacific Ocean, which lends itself to both this and other text, such as a legend. The maker can insert whatever they wish, but a cartouche usually includes the manufacturer's name and address and the globe's year of manufacture (though globes are very easy to date). The scale of the globe will also often be shown along with dedications or descriptive details specific to the globe, maker or purchaser.

Celestial Globe A spherical map of the major stars and eighty-eight constellations. It is mapped by a north and south pole, an equator, and lines of longitude and latitude in the same style as a terrestrial globe. It can be depicted one of two ways, as if the viewer is at the centre of the Earth, or as if the viewer is at infinity looking down upon the Earth, the imagery of the constellations varying according to the depiction.

Constellations When the heavens were first mapped, mythological creatures, gods, objects and animals were used to identify groupings of stars. These constellations are still an important aid to basic navigation.

Ecliptic The ecliptic is a plotted circle around the Earth that represents the apparent path of the sun among the stars throughout the year. It can also be defined as the plane of the Earth's orbit around the sun. Along with the marked meridian it allows the user of a globe to calculate where the sun will be overhead on a given day. During the year it will pass through each of the twelve constellations of the zodiac, which are found along the ecliptic with the moon and the planets, and which all lie in a similar plane to the Earth. This is because the solar system formed from a rotating disc of gas and dust surrounding the young sun.

Equator The imaginary circle that runs around the centre of the Earth, it divides the world into the two hemispheres. Because the Earth is tilted, the sun can be overhead up to 23.5 degrees north or south of the equator, but never beyond this angle. Because of the greater angle, the length of days varies much less nearer the equator than in higher latitudes, and sunsets can be over in a flash.

Equinoxes Twice a year the lengths of day and night are equal everywhere on Earth as the Earth's tilt is facing neither towards nor away from the sun, and is thus perpendicular. The dates, around 20 March and 23 September, are often seen to mark the beginning of spring and the end of summer.

Globe The three types of globe. A painted globe is simply a sphere that is painted to resemble the Earth. Even with an accurate image projection (onto a curved surface!), it is somewhat inaccurate. Detail is also very time-consuming to add. The same can be said for a manuscript globe, which only vaires in that has a paper layer added to the spherical ball, before the cartographic detail is added. A printed globe is by far the most accurate, especially with modern digital mapping software. The map is printed onto surfboard-like sections that affix to a sphere, and if done correctly this is the most accurate way to make a printed globe.

Gore One of the twelve, twenty-four, thirty-six or forty-eight segments of paper, printed with a map, that cover a printed globe. When using paper (as opposed to plastic shrink wrap or painted globes), the minimum number of gores will usually be twelve, due to its restricted elasticity.

Horizon Band Many, especially historical, globes are set within a table which has a circular wooden top, similar to Saturn's rings. To this globemakers attach a horizon band which, depending on size, may depict dates and months, the degrees and signs of the zodiac and points of the compass along with other astrological information.

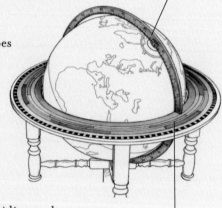

MERIDIAN

HORIZON
BAND

Hour Dial A thin metal ring, attached to the meridian and usually placed on top of a globe at the North Pole. It allows viewers to calculate the variation in time between the different time zones on the globe. Hour dials are usually not removable, which has the disadvantage of stopping the meridian and globe being moved through a full 360 degrees.

International Date Line (IDL) The line used as an international reference to determine the beginning and end of each day. This, along with many time zones, follows a convenient route rather than a straight line, to facilitate the functioning of business and for convenience between neighbouring countries. The IDL is on the 180th meridian, which is the only line of longitude that doesn't traverse a major land mass.

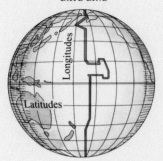

INTERNATIONAL
DATE LINE

Latitude Lines represented as horizontal circles on the globe marking the angular distance from the equator of a place on the Earth's surface. The equator is at zero or no latitude, and those above and below are referred to as north or south latitude. The poles are at 90 degrees. These lines allow simple referencing of arctic, tropical or temperate zones.

Legend Text on a globe to explain features shown, generally geographical and physical, but anything the globe depicts, including shipping routes, undersea cables, etc.

Longitude These are the vertical lines on a globe, and are depicted east or west of the prime meridian, which is found at 0 degrees, running through Greenwich, a borough in south-east London.

Meridian A meridian is the brass (usually) circular ring that surrounds a globe. Attached to the meridian is an axis which is clamped to the meridian at opposite sides, threads through the globe at both poles and supports the globe. The meridian is usually numbered from 0 degrees (at the equator) to 90 degrees at the poles.

Due to the globe often being set off centre, engraving can be on one or both sides.

Roller Bearings A bearing (ball transfer unit) consisting of a large main ball sitting on top of many smaller balls set within a sealed chamber. A precise number of smaller balls needs to be used so that the main ball neither jams nor spins over the smaller balls. Bellerby & Co. globes sit atop three main balls, which in turn sit on several hundred smaller balls.

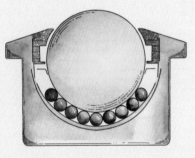

Solstices These are when the sun reaches its maximum northerly or southerly point (21 June and 21 December) relative to the celestial equator, and the dates in the calendar, dependent on hemisphere, which denote the longest or shortest day.

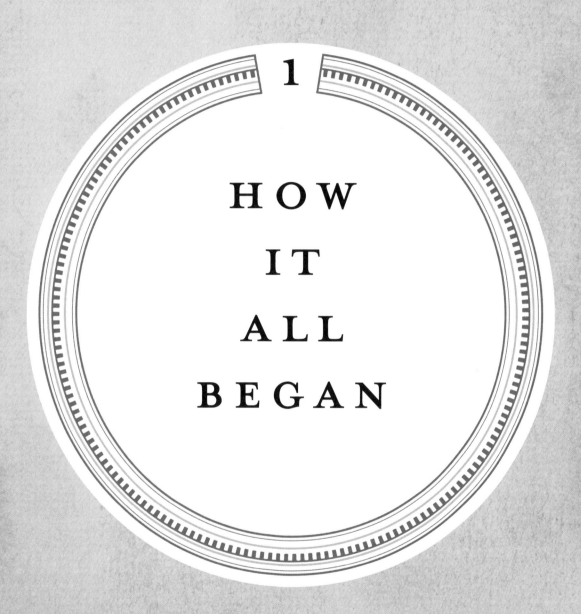

1

HOW
IT
ALL
BEGAN

Our planet is remarkable. I have always been intrigued by the notion that life on Earth is made possible by the coincidence amongst others of the Earth's distance from our star – the sun, the make-up of the atmosphere, its orbit around the sun, the length of its rotation and the precise angle of its tilt as it rotates, because if any one of these factors changed by a fraction, it would probably be a barren pile of desolate rocks, not unlike Mars.

'I WANTED TO UNDERSTAND HOW THE WORLD WORKS'

As a child I didn't read many novels; I had a vivid imagination but make-believe stories were not the medium to transport me. I was much more interested in facts. I would read any book on space, science and natural history. Growing up in the middle of nowhere, with my father working abroad for much of the year, I would spend hours in his vacant study, avoiding homework and poring over his collection of encyclopedias and illustrated books on the natural world. I wanted to understand how the world works. Like a lot of children, I was fascinated by the universe – and by extension, globes, the only accurate representation of our planet. I used to pester my parents to buy one of those garish 1970s numbers that you'd see in the Sunday supplements, which opened up to reveal a drinks cabinet, one of the few large floor-standing globes available at the time.

My interest in globes has been a constant ever since. I use Google Maps every day – to find my way to a new part of town or check something online – but it doesn't replace the apparently old-fashioned globe; for me the modern digital map and the globe perform completely different functions. You never use a globe today

Explore, Enjoy, Listen, Learn, Peace

for directions, and when you look at a map on your phone, you never experience the same awed feeling as you do when you hold a globe in your hand or spin it on its axis. Google Maps might inform, but a globe inspires. Perhaps this is because it gives you a different perspective from a flat map of the world. Whereas on many two-dimensional maps the focus seems to fall on Europe as the apparent centre of the world, the globe is a spherical object on which there can be no preordained centre, and thus each place is of equal significance. In that way, they help us to understand where we are in relation to the rest of the world. Globes remind us of how minuscule – and insignificant – we are. And how wonderful the world is, a beautiful planet floating in space, spinning within an infinite universe and an evolution of time so long that

he trained as a naval architect, working firstly at Cammell Laird on Merseyside, then at BP Shipping for more than twenty-five years. I thought a globe would be something he might really engage with. If I found the right one, it might even, I thought, make up for all the missed years. His birthday was a few months away, so I started looking.

First stop was Stanfords, the famous map shop, then on Long Acre in London's Covent Garden. They stocked plenty of brightly coloured classroom-quality globes, perfectly functional and not very expensive, but these were not going to work, not for a landmark birthday. There were a couple of businesses making globes in the UK and America, but either the quality and aesthetic was not what I had in mind, and I was left wondering how

'IN EARLY 2008 I DECIDED I WANTED TO BUY MY FATHER A GLOBE FOR HIS UPCOMING EIGHTIETH BIRTHDAY'

it is hard to comprehend. The concept of time is difficult enough to fathom when related to a 2,000-year-old giant sequoia tree, let alone a 4.5-billion-year-old planet.

In early 2008 I decided I wanted to buy my father a globe for his upcoming eightieth birthday. He was very traditional, not easy to buy presents for; he viewed them with suspicion, finding fault with most. Over the years, I had mainly bought him practical things like socks, ties and nice bottles of gin, but sometimes (quite often) nothing. It was simpler to be labelled forgetful or ungrateful than to buy him something unappreciated. After his mother had refused him permission to join Luton Town as a professional footballer in the 1940s, seeing it as a poor choice of profession (he later did the same to my elder brother),

anyone could make them so cheaply, or they produced not unattractive one-off faux-antique globes (along with other 'medieval' props) for movie sets that relied on ageing techniques (shoe polish, coloured varnish and tea) and soft-focus filming to disguise the lack of finesse. I then visited a few auction houses, where I watched the bidding for antique globes rise to tens of thousands of pounds, sheepishly hiding my paddle when I dropped out early on. I hadn't planned on spending that sort of money, but equally didn't see the point in having a globe that was decades if not centuries out of date and fragile to the point of being almost untouchable. These delicate antique artefacts belong in museums and stately homes. Besides, I knew my father, and he always favoured functionality over aesthetics. He had little interest in historical context.

Later that year, in June 2008, I resigned from my job at the central London ten-pin bowling venue I had helped a friend set up three years earlier (and subsequently been press-ganged into managing), packed my bags and set out on a backpacking trip with my partner Jade. We included several nations on the trip – India, Egypt, Morocco – where artisans working in small factories still flourish alongside street markets, bazaars and souks. At each stop on our six-month trip, we looked around in the hope that somewhere I might stumble on my perfect gift. In Marrakesh we visited small and busy artisanal workshops in simple houses

I found nothing suitable as a gift for my father. It was hard to believe that among the myriad of handicrafts we encountered on our travels, there seemed to be no one making beautiful modern globes by hand.

I had two options: a cheap, modern political globe (also available with a generous coat of sepia colouring and imaginatively labelled 'antique') or an expensive and very fragile actual antique. Hobson's choice.

Ever since I was quite young, I had one plan, and that was to work for myself. It is not that I don't like taking instruction (fact check: the truth is, maybe I don't like taking instruction) or being asked to do a

'THERE SEEMED TO BE NO ONE MAKING BEAUTIFUL MODERN GLOBES BY HAND'

with rudimentary foundries in back rooms. There we watched workers shape cast brass pieces by hand, terrifyingly, on what looked like hand-built, open-geared lathes, with razor-sharp shards flying in all directions. No gloves, no safety glasses in sight!

I was convinced that I would find something for Dad in India. On a previous trip around Rajasthan, I'd seen all manner of crafts in remote villages. Just outside Jodhpur we found the most amazing source of Indian antiques in Suncity Art, a space spanning the size of many football pitches. We spent a day there, buying several beautiful items but no globes. On the last leg, we ended up in Cairo. We spent hours wandering around the souks in Cairo's Khan el-Khalili mesmerised by the masses of hand-worked wooden chess boards, bone- and shell-inlaid boxes and other trinkets. But apart from a few simple antique brass celestial globes in Morocco,

task, I just don't like anyone telling me how to do the task. The most fun part of any job, by far, is working out the most efficient way to do it, though this won't necessarily be the quickest way first time round, as tests are necessary. And yes, I can see that in many professions this might be a very bad approach.

In 2001, aged thirty-six, I was made redundant from my position at ITV with a very generous severance deal and so was fortunate enough to finally get the chance to take control of my own destiny. I bought, single-handedly renovated and then sold a couple of houses over several years. While I made nothing on one property, I did well on the other, albeit due more to an upturn in the market than my handiwork. Along the way I had learned quite a bit about materials and machinery, which would later prove valuable.

THE ATLAS

A collection of maps has been called an atlas since the 1500s, when cartographer Gerard Mercator put a picture of Atlas holding up the Earth on the title page of his book.

In Greek mythology, the Titan Atlas, brother of Prometheus and Epimetheus, was the father of the Hesperides, the Pleiades, the Hyades and the nymph Calypso. He was the leader of the Titan rebellion against Zeus, the Olympian god of the sky and thunder. After the Olympians defeated the Titans in battle, as punishment Zeus condemned Atlas to stand for eternity, holding up the heavens (not the Earth as depicted in Mercator's book). According to the Greek poet Hesiod (c.750–650 BCE), Atlas stood at the western end of the Mediterranean, the edge of the known world. The ocean beyond was called the Sea of Atlas, or the Atlantic, in his honour.

Over time Atlas also became associated with north-west Africa, particularly Morocco. In 8 CE, in *Metamorphoses*, Ovid wrote that Atlas received a visit from Perseus, son of Zeus and slayer of the Gorgon Medusa, who asked for shelter. Because of a prophecy that a son of Zeus would one day steal the golden apples from his daughters, the Hesperides, Atlas turned him away. Insulted, Perseus showed him the severed head of Medusa, which had the power to turn all who looked at it into stone. Atlas too was turned into stone, and became the Atlas Mountains of Morocco.

Arriving back home after our travels in late 2008, I didn't have much on my agenda. Even with my total lack of financial acumen and general disregard for risk, I knew a financial crisis was not a good time to re-enter the property market, which had been my original plan. The economic crisis did however give me a little help. Bear with me on this. It's about mortgages so I'll keep it very short. In 2008 the Bank of England lowered interest rates to 0.5%. My mortgage repayments were calculated at 0.51% under this rate. I therefore found myself in the very odd position of my mortgage provider actually paying me to live in my own house. I was only making around fifty pounds a month and it was only for eight months, so I wasn't quite quids in, but in the short term the pressure to earn a living was off.

Before deciding on my next career move, I thought this was the perfect opportunity to devote some time to a little project, which I could do simply by increasing my overdraft. So, for the next few months, while I waited for some financial certainty to return, but without any consideration as to whether it was feasible, I decided to handmake a globe for Dad myself. Actually, I wanted to get it done so I could resume earning a living. I'd missed his actual birthday by now so

there was no rush, but even I could not have envisaged the scale of the task on which I was about to embark, nor the length of time it would take to complete.

I put my plan in action at a meeting with two friends in a King's Cross pub. One, Jon, I asked to do a simple bit of coding by writing a program to morph a flat map into gores (more on this later), and the other, Kelley, agreed to help part-time for a few months. I decided that I would make two globes, one for Dad, one for me. This increased to three when I realised the easiest way to pay Jon (the morpher) was to offer him a globe as recompense for his work. At this point I estimated it would take three to four months and cost a few thousand pounds. How difficult can it be to make a sphere and put a map on it, I reasoned? It never really crossed my mind why there might not be other globemakers doing everything by hand and that the process might be difficult. In truth, I was quite convinced that once I had made my globe, I would find other artisan makers around the world, but regardless it seemed like a fun project.

I began my research online. On YouTube there were some hilarious videos of people trying to make spheres, some passionate and rich (but time-poor) banking executives documenting all their work diligently and

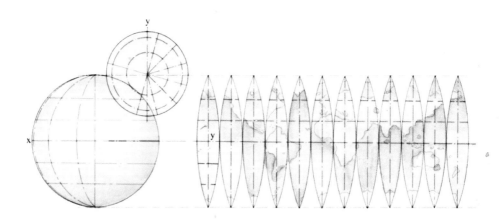

failing, and a few fantastic videos of two Italian men wearing white coats as if they were doing a major scientific experiment. I visited an enormous prop house in west London that had every piece of furniture imaginable, but no globes of note. It transpired that in the last sixty or so years, makers had stopped constructing globes properly; they'd been cutting corners (quite literally) and had turned the profession into a part-time do-it-yourself craft, lacking artistic refinement and technique. Some of the globes produced in the 1960s and 1970s did a better job at wiping out entire countries than many dictators. There were undoubtedly many reasons for this, although it might have been simply down to a lack of competition. But I didn't believe it was for lack of demand.

I began to think that maybe despite there being over seven billion people on the planet, there really was no one doing this craft to the level I wanted. Plainly, no one was using beautiful historical handmade and hand-painted globes as inspiration for their work. As for me, I had no fear of failure because at the outset I didn't have anything riding on it; I could always revert to buying Dad a bottle of whisky or some socks. But I was nevertheless confident that I could make something not dissimilar to the globes produced by historical artisans – notably, Blaeu and Coronelli, two of the most important makers technically and aesthetically (see p95 and p60), who had made globes so beautifully in the seventeenth century.

While it's inaccurate to say these makers took their construction secrets to their respective graves, I told myself that making a globe couldn't be complicated: a simple mix of basic maths (using pi), art and craftsmanship, all of which I felt I could master. I was confident I could work it all out and set about making a list of priorities.

1. License a digital map that I could edit. Learn how to edit using appropriate software.

2. Commission someone to write a program to morph a rectangular map into gores – the surfboard-shaped strips of paper, thicker at the equator and thinner at the poles, on which the sections of the world map are printed and which are then fitted onto a sphere. (Satisfyingly, I could cross this one out right away since Jon was already on board.)

3. Learn how to make plaster spheres.

4. Learn how to apply paper gores to a sphere.

5. Paint and varnish the globe – most likely find someone else to do this.

6. Find someone to help me construct the wooden bases on which globes sit.

7. Find a foundry to cast a brass meridian and an engraver to add the detail.

There would, of course, be other little mysteries I would have to work out along the way. How, for instance, do you balance the sphere, so that when it spins it comes naturally to rest? And how do you support the globe within the meridian? Engineering challenges: another reason to be excited about my foray into globemaking.

Arguably, I have a short attention span (I am after all on my fourth career, and that is to exclude the various starting-position

jobs in my twenties), but with this venture it was different. I was, shall we say, obsessed. As I became more engrossed in my mission to create the perfect globe, I basically went off radar; I didn't want anyone to think I was having a midlife crisis, trying to 'find myself' in some

something of note. No trumpet blowing, thank you. At the beginning, therefore, I just avoided the 'job' subject entirely, but as it became more serious, the money I ploughed in started to escalate and as the time went on and on, it was simpler to keep a low profile.

'IF I COULD TURN BACK THE CLOCK, I WOULD NOT CHANGE ANY OF IT'

weird art project. Jade didn't mince her words: she thought I was completely mad, although her mother was my biggest fan throughout. But I could just imagine the mockery I would receive from friends if I tried to present myself as a globemaker before I had even made a single globe. This is England after all. General MO: be quiet about any success pretty much always, but as a minimum until you have done

A year later, in summer 2010, my one-man operation had grown to the extent that I had hired a part-time painter and a woodworker, and it had evolved into my own company. I had completed a few globes, set up a website and as a result happily received a first order, from a librarian in Australia. Dad's birthday present, however, was going to be even later than planned – given the

precarious state of my finances, delivering a paid commission had to take precedence. With no proper records of expenditure, except by seeing the increasing size of my overdraft and credit card debt at the time, I have no idea how much it all cost, and probably do not want to know. And, if I had known how long it would take, how much money I would spend, how it would fracture relationships, I would probably have pulled the plug on the whole project. I certainly would not have started something with so much uncertainty about the outcome. One thing is for sure: had I stopped there, my first commissioned globe, now sitting in Brisbane, would have felt like one of the most expensive in history.

For the first five or six years, until well into 2014, sales were barely covering costs and I was therefore still taking no salary. I had no budget to promote our product and no established market for bespoke handmade globes to present ourselves into. To fund the project, I also had to sell my car and shared house in Stoke Newington to pay down some more debt. It seemed to be a bottomless pit. And yet, if I could turn back the clock, I would not change any of it.

What had started as a modest plan to make a special birthday gift for my father led me to a whole other life as the founder of Bellerby & Co. Globemakers, the only fully bespoke globemaker in the world. We do everything from start to watercolour finish by hand, and as a result I find myself more in awe of our planet than ever. Globes are a wonderful convergence of the arts and the sciences. The process of globemaking requires sound knowledge and skill in engineering, geographical knowledge, artistic ability in painting and, most importantly perhaps, an endless curiosity. The job of a globemaker is never complete: politically, the world changes continuously, and the act and art of globemaking is just as captivating as the countries, cities, mountains and oceans depicted on the globes themselves.

2

THE
ROUND
EARTH
ROLLS

UNDERSTANDING OUR PLACE IN THE INFINITE UNIVERSE

People have been fascinated by the idea of replicating our planet for many reasons, but perhaps most compellingly of all because there is nothing like a globe to make us really think about our place in the universe. Many of us first encounter a globe as children at school or, if we are lucky, at home. As we spin the globe, our parents or teachers may explain that our home city or town is but a negligible spot on a huge sphere, which is itself tiny and dwarfed by the galaxy and in turn the infinite universe beyond. Gazing at images of our planet, seemingly suspended in a massive void among the stars, there are so many questions that need answering, many of which require a level of knowledge way beyond the comprehension of a young child. Gravity and infinity, for example, are certainly neither easy to explain nor quickly understood even for a grown-up.

The scholars of the sixth century BCE were the first to figure out that the Earth was spherical (which obviously asks questions of certain contemporary theorists). In the fourth century BCE the Greek philosopher Plato travelled to Italy to study Pythagorean mathematics. Returning to Athens, he established a school where he taught his students that the Earth was a sphere. If one were able to view the planet from up in the clouds, he wrote in his *Phaedo* of 380 BCE, it might resemble 'one of those balls which have leather coverings in twelve pieces, and is decked with various colours, of which the colours used by painters on Earth are in a manner samples'.

Aristotle (384–322 BCE), Plato's star pupil, took his investigations further. Observing that 'there are stars seen in Egypt and [...] Cyprus which are not seen in the northerly regions', he set out to find conclusive physical proof that the Earth was round. Later he noted that as it sailed over the horizon, a ship appeared to disappear hull first, indicating that the surface of the Earth is curved. Looking at the moon during a lunar eclipse, Aristotle also observed that the Earth cast a round shadow on its surface.

this time in Syene the angle of the sun in relation to an identical vertical stick was zero degrees, Eratosthenes concluded that the sun's rays were parallel, while the Earth was curved. Using the estimated distance from Alexandria to Syene and the difference in angle between the two cities, the mighty Eratosthenes could then calculate the approximate circumference of the Earth, which he did to within 2 per cent of the accepted modern value of 40,075 kilometres (24,901 miles). However, this figure

'THERE ARE SO MANY QUESTIONS THAT NEED ANSWERING'

A century later, while working at the Library of Alexandria in Egypt around 240 BCE, the Greek astronomer Eratosthenes of Cyrene was the first to calculate the Earth's circumference. A correspondent from the city of Syene (now Aswan) in southern Egypt informed Eratosthenes that, at noon on the summer solstice, the shadow of someone looking down a deep well would block the reflection of the sun in the water at the bottom. In other words, on that date, in that geographical location and at that precise time, the sun was directly overhead. However, Eratosthenes knew that on the summer solstice in Alexandria this was not the case; where he was, objects still cast shadows.

At midday on the solstice in Alexandria, Eratosthenes measured the length of the shadow cast by a vertical stick. He then calculated the angle the sun made with the stick. This was one-fiftieth of a circle, or 7.2 degrees. Knowing that at

was not widely adopted by Eratosthenes' successors, with significant consequences for the development of cartography.

Who then made the first globe representing the Earth? No one can say for sure, although some sources credit Greek philosopher and grammarian Crates of Mallus with envisaging a globe bearing a map of the known inhabited world in about 150 BCE. In his *Geographica*, published in 7 BCE, the Greek philosopher, historian and geographer Strabo suggested that for such a globe 'to present the proper appearance to observers', it would need to be no less than ten feet in diameter, but there is no evidence that Crates actually constructed one himself. And while there are surviving examples of celestial globes, made with punched and hammered brass or silver, from as early as the second century CE, sadly no terrestrial globes survive from this time. For the ancient Greeks, mapping the stars was, let's face it, a little easier than exploring the planet.

BEHAIM'S GLOBE: THE ERDAPFEL

Whilst celestial imagery and depictions date back to ancient times, we must jump forward to the late fifteenth century to find the world's oldest surviving terrestrial globe, known as the Erdapfel – literally 'Earth apple'– made between 1492 and 1494 by Martin Behaim in Nuremberg, Germany. It is one of the last important pre-Columbian representations of the world.

Martin Behaim, an ambitious cloth trader, came from a prosperous and prominent Nuremberg merchant family who sent him to Portugal as a young man to gain experience in business. There he seems to have spent much of his time among the sailors and navigators around Lisbon's busy docks and boasted of travelling widely himself, supposedly sailing as far as the west coast of Africa. The young Behaim also gained an introduction to the court of King John II of Portugal, where he supposedly found favour as a counsellor, advising on astronomy and navigation.

Returning to his native Nuremberg in 1490, Behaim was keen to pass on his knowledge of the new maritime trade routes being opened up by the Portuguese to the city's councillors, who in early 1492 commissioned him to make 'a printed mappa mundi embracing the whole world'.

Perhaps he recognised that a globe was the ideal way to demonstrate how easily a merchant could reach the Indies and its promised riches by sailing west, and so the idea for the Erdapfel was born. Back in Nuremberg, Behaim enlisted the help of a team of prominent craftsmen to construct a sphere roughly twenty inches in diameter. They pasted linen strips onto a clay ball, then when these had dried and hardened, cut the linen at the equator to free two hemispheres from the ball. These were strengthened with wood then joined together with more paste. Behaim drafted his map, evidently basing it on an existing world map, and amended it using information from expeditions by

THE EARTH - SOME STATISTICS

Land makes up around 29% of the Earth's surface (148,940,000 sq km). Asia is both the largest and the most populated continent at 44,579,000 sq km.

At the time of writing, the world population stands at 8 billion, having doubled in the past 40 years, and is on track to be nearly 10 billion by 2050. This averages out at 60 inhabitants per square kilometre, the highest density being found in Macau at over 18,500 people per sq km, while less populated places like Mongolia have fewer than two people per sq km.

Marco Polo and more recent findings by Portuguese explorers. It was then painted onto the surface of the globe in exquisite detail by Georg Glockendon, a Nuremberg woodblock printer and painter of miniatures.

On the globe, the Americas are absent, as Christopher Columbus was still on his travels (his fleet made land in the Bahamas in October 1492, only returning in early March 1493); the Eurasian continent is oversized, and there is a huge ocean between Europe and Asia, with Japan (Cipango) out of position (too far south) and enormous. The globe includes the mythical St Brendan's Island in the Atlantic. There is also a gigantic phantom peninsula

the Behaim family. In 1907, after two cack-handed attempts at restoration in the nineteenth century, it was transferred to the German National Museum in Nuremberg. In 1992 it went to the Vienna University of Technology, where it was studied in high resolution by the Behaim Digital Globe Project. Further digitalisation, by the German National Museum, took place in 2011. It is interesting to note that the colloquial translation for Erdapfel is 'potato'. Potatoes are native to North and South America, and when Behaim made his globe, they were a treat as yet unknown to Europe.

'ITS RESEMBLANCE TO THE WORLD WE KNOW IS QUITE REMARKABLE'

to the east of the Golden Chersonese (Malay peninsula). Overall however, considering the constraints on exploring, mapping and positioning in the late fifteenth century and the rudimentary cartographic tools with which Martin Behaim created the map on his globe, its resemblance to the world we know is quite remarkable. But it is also highly inaccurate, in places based on little more than educated guesses with many areas simply unknown.

More reliable perhaps were the other pieces of information Behaim included on his globe: he listed the commodities to be found in different regions, the location of markets and local trading protocols, all of which would have been invaluable to his German mercantile sponsors.

Until early in the sixteenth century, the Erdapfel was kept at Nuremberg town hall, after which it was kept in storage by

The globe has a great deal of additional detail now badly darkened by age, including over one hundred miniature objects and figures – flags, saints, kings on their thrones, elephants, camels, parrots and fish, fantastic creatures including a sea serpent and a mermaid – so as well as being the oldest globe known it also gives a snapshot of European knowledge at the time.

One interesting question is, why did Behaim only make one globe? Well, before 1500 and widespread access to printing presses, all globes made were manuscript globes: in other words, unique and often bespoke items, on which artists engraved, drew or painted a map entirely by hand. This made globes such as the Erdapfel painstakingly slow and costly to make, one-offs made for a particular person or function, which the globemaker could not reproduce to order in quantity.

MAPS AND GLOBES

Maps are hugely important tools in our everyday life, whether as practical guides helping us find our way from A to B, or inspiring and forming our ideas about the universe.

Today, our hefty world atlases, city A–Zs and AA road maps have largely been supplanted by handy mobile phone apps and GPS screens in our cars, yet for many of us there is nothing more magical than losing ourselves in a good old-fashioned paper map.

Similarly, globes are spherical maps of the Earth or heavens, beguiling objects full of detail, colour and wonder. They are especially useful because their rounded shape shows a planet without distortion. But in the past, they were much more than this. For the Ancients and later for Renaissance students and scholars, used as a pair, globes allowed them to demonstrate and investigate the relationship between the heavens and the Earth. For early marine navigators, the engraved brass meridian rings and the printed horizon rings encircling old celestial and terrestrial globes were not mere decorative features, but helped to plan a course at sea. For ambitious merchants and expansionist rulers, they helped plot new trading routes and colonial conquests. With the advent of the mechanical printing press in the 15th century and the subsequent proliferation of printed globes, they played a key role in the distribution of new knowledge about our world. And then as now, they neatly embodied the mystery of existence and the need to find our place in the cosmos.

Today, thanks to space travel and satellite technology, the terrestrial and celestial globe has been joined by the planetary globe, with globemakers, including Bellerby & Co., crafting accurate representations of the surface of the Moon or of Mars.

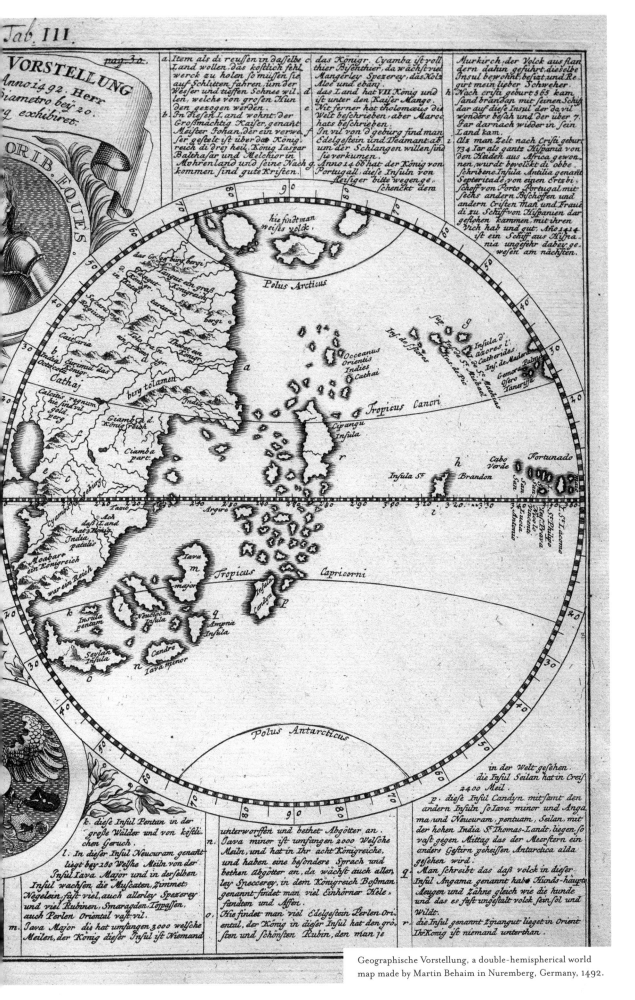

Geographische Vorstellung, a double-hemispherical world map made by Martin Behaim in Nuremberg, Germany, 1492.

THE SECOND AGE OF EXPLORATION

Following the initial voyages of discovery in the late fifteenth and sixteenth centuries, the eighteenth century is often called the 'second age of exploration'. After navigational errors led to the wreck of four Royal Navy ships off the Scilly Isles in 1707, the British Government established the Commissioners for the Discovery of the Longitude at Sea (also known as the Board of Longitude), offering large amounts of money to anyone who could come up with a method for accurately establishing longitude. This motivated a more practical and scientific system for navigating at sea in the eighteenth century.

For a new generation of rationalists, globes seemed arcane alongside their more modern scientific apparatus, such as air pumps and microscopes, and new-fangled sextants and chronometers. A new breed of astronomers and navigators maintained that globes might well be useful to record new discoveries, but not to make them. Thus, increasingly, the use of the celestial globe as a scientific tool and an aid to navigation fell out of fashion.

Terrestrial globes continued however to prove very useful for mapping the new discoveries made in the eighteenth century, including many of the Pacific islands and New Caledonia. Globemakers constantly updated the known existing territories on their maps in line with continuing exploratory ventures, so globes remained invaluable for merchants and traders, plotting new routes, and as educational tools for rapidly and concisely spreading knowledge of geographical and astronomical discoveries.

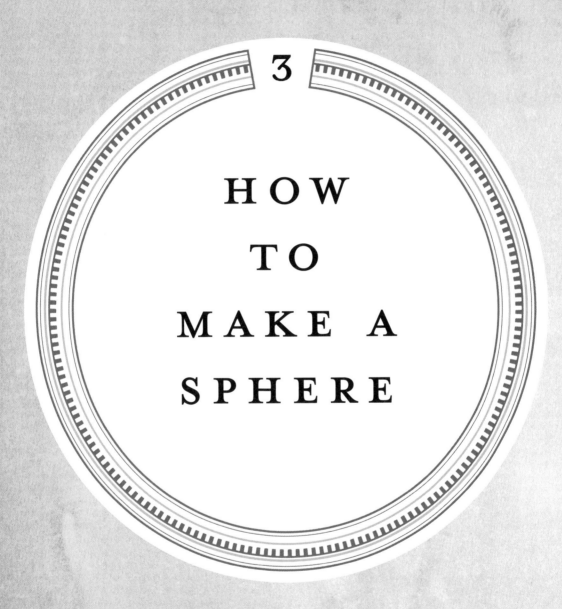

3

HOW TO MAKE A SPHERE

In the modern digital age, we can now print almost anything to scale in 3D in numerous different materials, but hundreds of years ago, craftsmen had much cruder equipment and measuring tools. As I learned the hard way, the frustrating thing about making spheres is that they are annoyingly hard to measure, and this difficulty only increases as they get larger. Which also means that, without very expensive scanning software and equipment, it is difficult to tell how accurate any sphere or mould you've made – or indeed that anyone else has made for you – actually is.

'HUNDREDS OF YEARS AGO, CRAFTSMEN HAD MUCH CRUDER EQUIPMENT'

My point of reference, howeve – the Earth – is not quite round; it is often described as an oblate spheroid. Compared to its girth at the equator, the Earth's circumference is a little shorter around the poles – in other words, it is flattened or depressed at the top and the bottom. This is due to the Earth expanding around the equator because of the centrifugal force created by its rotation. But, if you were looking at the Earth from an interplanetary spaceship, this would not be apparent to the naked eye.

From the outset, making a sphere was the main focus of my time, energy and learning. I was starting from scratch, creating and discovering my own techniques and work-throughs, so whatever I did would have repercussions and consequences going forward. There is, I discovered, a sound reason most building structures are essentially rectangular. With a square side, you can monitor

progress and check accuracy using a simple laser pointer, or even just the naked eye, as your construction develops. If something does go wrong, it's easy to spot and usually easy to fix. When making a sphere, accuracy is paramount, as the form accentuates any and every issue. It's as if any error is multiplied by pi (approximately 3.1415), and not only is it difficult to check that everything is going to plan during construction, but also if something does go wrong, if any measurements are off, it is not fixed easily, if at all.

I decided to make a globe fifty centimetres in diameter. With the benefit of hindsight this was probably far too big as a starting point, but I wanted to give my father a globe with gravitas. At this stage I had no way of knowing the complexity of the task I had set myself or of figuring out what difference just a few centimetres would make.

I still had to decide whether to make my sphere oblate or perfectly spherical. Given that making a perfect sphere was going to be challenging enough, and as I later realised, even more difficult as the diameter increases, should I further complicate

matters by flattening the poles? I came across a theory online that if you were to shrink the Earth to the size of a snooker ball, it would fit within the regularity specifications of an officially approved ball. I like this analogy: it brings clarity to the accuracy of topographic globes; it appears factually correct, in that the Earth would be as smooth, but is not quite round enough due to said equatorial bulge. The ball would therefore struggle to follow a straight path.

Moreover, it seemed obsessive to focus on something that is difficult to visually affirm or negate. Given the difficulty in measuring a sphere, how could I then gauge its relative oblateness? And how do you make it oblate in the first place? While globemakers strive to make perfect representations of the Earth, they are still artistic creations. Other considerations take precedence over replicating something not visible to the human eye, and while I was an early adopter of computers, I've always avoided being unnecessarily techie. So I decided to aim for a perfect sphere; at the scale at which I was working, the globe would still be incredibly accurate, and enough of a challenge as it was.

PRECESSION

Every 26,000 years, the orientation of the Earth's axial direction makes one revolution around its rotational axis. Picture that the Earth's axis is now pointing to the left – the Earth tilts at 23 degrees, so assuming we hold north as the top position, it needs to face one way – then slowly over thousands of years that tilt makes a full revolution. Imagine it as the movement of a spinning top as it slows – while the top is spinning fast around its axis, the angle of that axis is also slowly revolving as if to paint a circle above it.

This slow change in the direction of the Earth's rotational axis is called 'precession', which is also why when plotted over the centuries, the stars appear to change position in the night sky, and these changes can be called 'epoch's'.

With further procrastination becoming just blatant time-wasting, I tasked myself with making a basic sphere. From the start I wanted to utilise plaster of Paris (now we use modern composites and resins). Next, I needed to settle on my method. At first glance, it may seem that using a mould to form two half-spheres and then attaching them together to form a whole was the simplest solution, but this required an understanding of mould construction, engineering, materials and skilful craftsmanship.

There were easier, if more crafty, alternatives. One option used by past globemakers was to coat a papier mâché sphere on an axle with plaster of Paris, while simultaneously spinning it through a semicircular template to smooth the surface. Messy, great fun to film for videos and handy for using up old newspapers, but not so practical for daily production! Another was to spin plaster of Paris into a preformed spherical mould, leave it to set and then unbolt the mould. But this would

SPEED OF THE EARTH

If you are standing at the equator, you will be moving at around 1,037 miles per hour (0.5 km per second); while simultaneously moving around the sun at 67,000 miles per hour (30 km per second); while simultaneously, along with the sun and all the planets, spinning around the centre of our galaxy at 490,000 miles per hour (220 km per second). Quite fast. Despite these speeds it takes around 200 million years for one full galaxy orbit. On top of that our entire galaxy is moving through the universe at 1.3 million miles per hour (581 km per second).

We are headed for the Virgo cluster, currently about 65 million light years away. Speed, time and distance are ludicrous on a galactic scale. The idea that we should prepare for any alien contact is absurd. I'm not sure we should look to Hollywood for an accurate portrayal of our first alien contact. If they even deem us worthwhile, anyone from a distant planet who gets here will no doubt extract all our minerals, resources and life in the space of a breath.

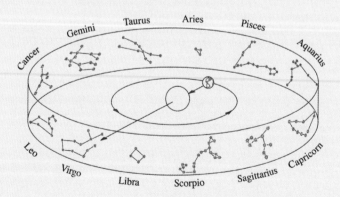

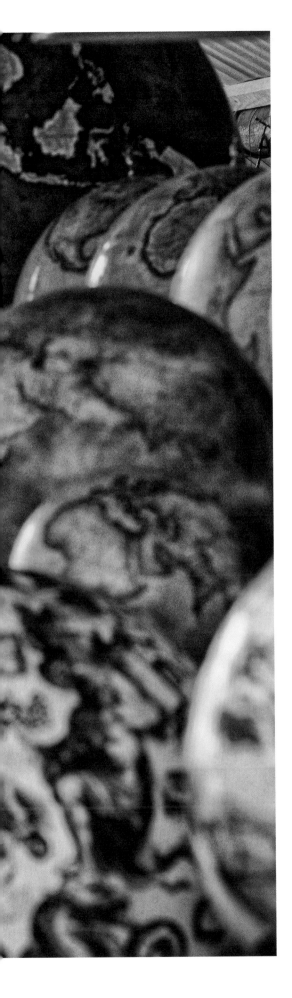

have involved me engaging the services of a specialist company, and as I only wanted to commission three pieces, the costs would have been prohibitive. Besides, I wanted to do as much as possible myself, so I reasoned that the two-half-sphere method was the most accurate and practical.

I somewhat naively thought I could rely on a manufacturer to make a perfect half-spherical mould from which I could cast the hemispheres, so I searched online for mould makers, quickly finding out that it's a niche industry in the UK. I managed to find several companies ('several' might give a clue as to how well this progressed) prepared to make moulds at the agreed fifty-centimetre diameter. At least that is what I asked them to do, but they kept mentioning 'tolerance', which I sort of ignored. To you and me, or at least anyone outside engineering, 'tolerance' is best described as error or, more technically, margin of error. But in the manufacturing business, as I discovered, the term disguises the fact that perfection is often a percentage game. Tolerance is the margin between what the customer asks for and how close the engineer thinks they can get to fulfilling their request, often without the customer noticing. However, when I was ordering my mould, it felt like I was in a game of hide-and-seek with the truth, except no one had told me I was playing one. Obviously, I had other measurements that were dependent on the accuracy of the sphere, hence my need for clarity.

GLOBES IN THE NEW WORLD

Towards the end of the eighteenth century in the United States, as in Europe, geography and astronomy were part of mainstream education, and globes and maps were in demand as teaching tools. These were first imported from England, but by 1810, James Wilson (1763–1855) of Vermont, a self-taught globemaker with an agricultural background, had established himself as the first commercial manufacturer in America. Others, including Bostonians William Annin and Dwight Holbrook, a maker of general school apparatus, soon followed in his wake, and Boston rapidly emerged as the centre of American globemaking.

This all changed with the expansion of lithographic printing in the mid-nineteenth century when Chicago took on the mantle of headquarters of cartographic publishing in the United States, and globemakers quickly followed.

A. H. Andrews learned his trade at the Holbrook family firm in Boston where he worked as a clerk, and set himself up in Chicago around 1860 as a supplier of office and school furniture and educational tools, including globes. Andrews marketed many of his first globes as 'Holbrook models' and later diversified, making a wide variety of fine editions in a range of sizes, from a small 3-inch pocket globes and compact desk globes to large 30-inch (76.2 cm) globes with ornate metal mounts. When A. H. Andrews Co. went into bankruptcy in the late 1890s, C. F. Weber, originally of Los Angeles, took over the business, later joining forces with T. H. Costello to become Weber Costello in 1907. This became the leading cartographic firm in the United States, and Weber Costello continued its globemaking through to the early 1960s, producing the famous globes for Churchill and Roosevelt in the Second World War.

Using newer printing techniques, Chicago printers and publishers Rand McNally began map and globe production in 1880; in 1930, Luther Replogle began selling globes which he made by hand in his Chicago apartment, using maps from the UK, and Replogle Globes was born; the George F. Cram Company of cartographers, established in 1869, entered the globemaking market in 1932.

As modern printing methods including photo engraving appeared, this boosted the Chicago globemakers' accuracy and efficiency, although it undoubtedly also made the end product increasibly more standardised.

Finding an accurate mould was an introduction not only to 'tolerance' but also to UK manufacturing and construction. It was a dispiriting experience. My requirement at the outset was neither comprehensive nor large (it still isn't), since we do so much in-house. Far too many people turned their backs when I asked a difficult question or mentioned a low production quantity; the immediate return was not large enough and any long-term profit was either too distant or not obvious. Over time I commissioned moulds from several companies, all of whom wanted payment in advance irrespective of whether the product they supplied me with was fit for use or even done as requested.

The first batch I received were in beautiful, hard, clear Perspex. Full of enthusiasm I cast the first hemisphere. It set quite well, but almost immediately I could see there was a problem. The mould

was a smidgen over a hemisphere, but still maintained its spherical integrity, and so by definition any mould would have a larger diameter than that of the point of extraction. So, its widest diameter was a little below the slightly smaller entry point. Even if I could release the cast, it was impossible to remove it from the mould in one piece. The next, from a one-man operation, was not much better, with flat areas all over the moulds. Upon querying ('Why are these moulds so bad?'), I was told that they had been vacuum formed and that he usually catered for a more niche craft clientele – hence the lack of concern about minor irregularities.

When I ventured an hour or more north of London, however, I found some manufacturers who approached the challenges I posed with open minds and willing hearts. Today our limited procurement is almost all with companies based in the northern counties of England.

THE EARTH'S ROTATION

The Earth rotates around its axis in 23 hours, 56 minutes and 4 seconds (known as a sidereal day), but it has also moved position in its orbit around the sun, so a further 4 minutes puts it back in its position relative to the sun (a solar day). This is also approximately 361 degrees of rotation. The shorter sidereal day refers to a full axial rotation when measured against the stars and other 'stationary' objects.

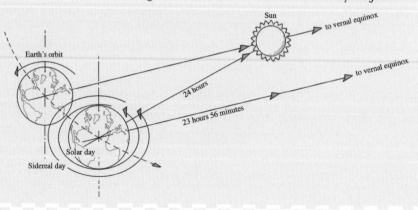

Then there was the problem of my workspace. In 2007 I had just finished remodelling our house, finally removing the bomb shelter from the back garden after sixty-five years of underuse. We now had a large open-plan kitchen, diner and living room on the ground floor with open steps leading to the upper floors. This was where I made my first attempts at creating a sphere, taping thin plastic sheeting floor to ceiling to mask off the rest of the house.

Plaster dust is so light it hangs in the air and the slightest wind moves it. Within seconds of starting work on the casts, white dust mushroomed everywhere. Every surface was coated. We had two beautiful cats, Randolph and Mortimer. Cats, as their owners know, have a habit of refusing to understand why somewhere is inaccessible. Show them a closed door and they will sit there and complain. Open it and they will just sit there. If you leave it open and disappear, they go back and forth like yo-yos, but you would never know as they are back sitting in their original position when you return. Randolph and Mortimer found the access door I had made in the sheeting, then made sure they stopped for a few seconds each time they went through it so their tails held it open and plaster dust billowed through. Even without the cats it would have been almost impossible to control. When choosing the stair carpet, I had suffered some sort of meltdown and plumped for a deep crimson. For six months I hoovered that carpet three or four times a day and wiped down every surface, every skirting board. If I did not keep to this schedule the cats would make me pay with paw prints every where. Morty had also recently had an operation to remove his pituitary gland to (as the hospital somewhat optimistically believed) cure diabetes. As we discovered after he returned from an unscheduled extra ten days of intensive care, a side effect was that he held his head at a ten-degree angle. Watching him watching me kept me sane!

EXTRACTION

Like most children, I had done some simple casting using plaster of Paris at school, but I did not remember the part about how to extract the hardened cast from the mould. I did more online research. One idea was using Vaseline or similar to grease the inside of the mould, just like buttering a cake tin in baking. But this was not a static or solid moulding; the process involved continually brushing liquid plaster over the curved surface as it set in place, which was a matter of seconds – any other liquid in there would therefore get mixed in with the plaster. I tried the Vaseline anyway. Surprise, surprise, it blended with the plaster, which I had to assume reduced the integrity of the cast. Besides, it didn't help extraction.

In another method, I hacksawed a few inches up one side of the mould from the widest point, then taped over the join. This, I imagined, would allow me, once the plaster had set, to undo the tape and prize open the mould to gain some traction on the cast. However, as the plaster set it would heat and expand, ripping open the tape and mould. Each mould and cast thus became progressively larger – obviously not ideal when you plan to stick two supposedly identical halves together.

My research kept pointing me to compressed-air extraction being a good method. If I could blow air in through a hole in the top of the mould this might work, but the examples I looked at didn't cover how to extract hemispherical forms, but other more easily removable shapes, often with soft silicone casts that simply peeled off. And I didn't have a compressor – at this stage my toolkit consisted of a cheap yellow drill – and I was reluctant to spend the money to buy one. I only planned to make three globes and didn't want to end up with a house full

THE HEMISPHERES

The majority of the Earth's habitable land mass is located in the northern hemisphere, and 80% of the southern hemisphere is water. Consequently, approximately 90% of the world's population live in the north.

of expensive machinery, although as the broken casts piled up, seemingly by default I was beginning to think that this might turn into a business. For the moment, though, I passed on the compressor: too expensive and I could find little evidence that it would work for me.

I persevered with my idea of cutting small channels up the side of the mould and then reinforcing the mould so that it stayed put. It seemed to work to begin with but collateral damage was high. Barely any casts survived extraction in a usable state; the moulds were degrading quickly, and even when I got them out in one piece the casts did not exactly inspire confidence. I had by now worked out a method using hessian to reinforce them, but it also made them harder to extract.

It was probably at this stage, mid-2009,

that I realised that my endeavours could become very expensive. (They already had!) It was obvious that as much as I loved my dad, I wasn't going to spend what would turn out to be nearly £180,000 just to make him a globe. He was lovely, but not that lovely. I decided that I would try to turn the globemaking project into a business. I had to bite the bullet, so I bought a compressor, which cost me several hundred pounds. Not only that, but its job was essentially over in a flash – all that money for five seconds' work!

It took several attempts using different methods before I found one that worked. I realised it was critical that no air escaped between the pilot hole that I had drilled into the top of the hard mould and the gun attachment on the compressor (requiring the purchase of a 10p rubber washer). I held the gun firmly against the hole. There was an immediate build-up of pressure as the air surrounded the mould. I held my breath. It felt as if the compressor or mould was about to explode, like the brain surge you experience when blowing up a balloon, then there was a loud bang and the perfectly formed hemisphere slowly released itself as if to say, 'What's all the fuss about?'

In autumn 2009, fed up with the stair carpet turning pink each day and the endless plaster dust, Jade told me to find a studio. Another unavoidable expense. Not far away in a mews called Leswin Place I found a room to rent next door to a nightclub, the fancifully named Stoke Newington International Airport.

Eventually I managed to improve the casting and extraction method to the point where most casts survived, and I could progress. Now I just had to join the two plaster halves together to form a sphere.

The extracted halves had rough edges, from which I needed to accurately trim off around an inch. Unsurprisingly, I could find no tools to measure accurately, cut and create spherical (or half-spherical) objects, so this was the beginning of a journey into designing and making my own bespoke apparatus and tooling, which was fun because it really felt like I was doing my own project and was especially rewarding. Cutting through a thin hard plaster cast was anything but fun.

First, I tried using a hacksaw. This worked to a degree but was painfully slow, so I tried a wood saw. If I could keep the saw moving it was fine, but if a tooth caught on the hard edge (extremely likely), it put the plaster under too much stress, leading to chipping and ultimately structural failure. Few casts survived and none were usable. Next, I turned to power tools. Despite my (quite rational) fear of angle grinders, I set about the plaster with an open metal blade spinning at thousands of revolutions per minute.

At first there was a horrific whine like a thousand piglets squealing as the blade cut through a material not much softer than stone, shortly followed by a tornado of plaster dust that rivalled a Saharan sandstorm in choking intensity. A fragile hemisphere, flexing under its own weight – fifty centimetres across and just two or three millimetres thick – is not the easiest thing to support, so the process involved lots of stopping and starting to readjust the cutting angle and position, and each time I restarted, the cut line almost instantly became invisible. Between each burst of the angle grinder, I had to wait several minutes before I could see more than a foot in front of me, and each time I ended up completely

My next machine was a Dremel multi tool – a cheap American DIY device which comes with various attachments for cutting, sawing, sanding, grinding and polishing. I was optimistic as the online demo videos showed a more professional model than the one I remembered. And amazingly for a home tool, it did quite a good job. Again, the Dremel was hard on the ears, but with it I achieved a relatively accurate and smooth cut. The problem was pace. It lacked torque, so it was a tediously slow process. To trim one hemisphere took over an hour – not an efficient solution.

I had now set up as a business, initially as a sole trader (later in 2011 as Bellerby & Co. Limited), and had created a rudimentary

'FEW CASTS SURVIVED AND NONE WERE USABLE. NEXT, I TURNED TO POWER TOOLS'

covered in fine white powder. While this method worked, it was just too aggressive and dangerous. I had to try alternatives.

My new workspace was a room in a shared studio. I had expertly gaffer-taped a bathroom extraction fan onto the window, sealed the door with more gaffer tape during the grinding operation and did my best to clean up afterwards with a vacuum cleaner, but despite my best efforts, the dust clouds were so extreme that there was little I could do to mitigate the mess inside, while the dust extracted with the fan just blew around the side of the building and in through the front door. Even when the room cleared and the dust settled, the slightest gust of wind would send clouds pouring out into the main studio. I held my breath.

website. Consequently, I had (astonishingly) taken my first order, from a librarian in Australia. I was also in the process of doing some building work on my house, and my Polish builder, Pavel, and his team were fantastic – diligent, hard-working, no tea breaks and long hours, keen to work six or even seven days a week, and non-complaining. It was like one of those situations where you can't quite believe that it's happening, but you don't want to say anything in case it's all a dream.

I've always believed that whatever you do, at whatever level, you should endeavour to improve on what has been done before, or at least do the best job you can, otherwise what's the point? I've also always thought that if you put in extra time and effort, people will notice. Not only were Pavel and his team good at what they did, they enjoyed their work and frequently worked so late into the night that I had to ban the use of power tools for fear of upsetting the neighbours.

I decided to offer Pavel a job as a freelance woodworker, initially to help me make some wooden bases, but also to act as a sounding box and to offer construction ideas and advice. I described what I wanted, and he suggested building a jig for the bandsaw, one of several more power tools I had recently purchased. I had experimented with the bandsaw on the plaster hemispheres; it was not only aggressive and incredibly loud, but it required holding my hands so close to the blade that I said a little prayer each time I used it. The jig he designed and built meant I could rest the hemisphere on a platform, keeping it steady and enabling me to guide the saw better. I still wore the thickest leather builders' gloves I could find, but it was much safer to work with and, lo and behold, it worked. The resulting dust clouds, however, were still extreme, so I had to look for a new studio. Or more accurately, I was asked to look for one.

EQUATORIAL BULGE

The Earth and most planets are larger around their equator than around their poles. This is simply down to centrifugal force. As a planet spins it will usually form a bulge at its equator. The extent of this bulge will depend on the speed of rotation and the density of the planet. In our solar system, Saturn has the greatest bulge, a noticeable 11% larger around the equator than its poles. This is due to a rotation period of just 10.5 hours, and a mean density that's less than water. Earth has a 0.3% difference.

Shortly after the formation of the Earth, it almost certainly had a much larger bulge as its period of rotation was around 6.5 hours. The Earth's rotation slows by around 1.4 milliseconds per solar day per century (which adds up over 4.5 billion years).

A further effect of the rotation of the Earth is that the speed of rotation at the poles is almost zero, while the speed at the equator is approximately 1,670 km per hour (1,037 mph), and is simply down to the larger distance that needs to be travelled. This results in the gravitational pull at the equator being weaker. This along with slightly less momentum of the Earth makes equatorial regions ideal, or at least preferable, places for space launches, so many sites are located close to the equator.

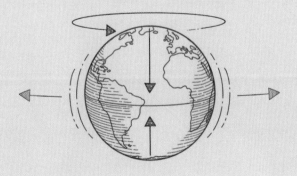

In early 2010 I moved to a workspace in Kynaston Road, just south of Stoke Newington Church Street in north-east London. It was a little shop on a quiet street owned by Joseph, a lovely Hasidic Jewish man from Stamford Hill, London's answer to Borough Park, Brooklyn. It had running water, three rooms, a yard and my favourite, following a great British tradition, an outside toilet. On top of that there was a huge window frontage. This would later prove useful, and at £600 a month the rent wasn't bad. So, there I was, still trying to make a sphere, but now I had a year's lease on a shop, one firm order, and time was pressing. Jade and I spent a weekend doing up the shop. We painted the interior the darkest colour available to cover up the state of the building, then I set to it. I also started to receive letters from the council. Which I ignored. 'Do not engage, do not respond' was my motto where Hackney Council was concerned. A few months later the bailiffs arrived, looking like East End gangsters in knee-length leather coats and refusing to believe I had never heard of business rates. Pavel let them in, and they refused to leave until I had paid up. Quite a performance for three months' arrears at £60 per month.

The process of fixing the two hemispheres together was something that took many trials to perfect. I made more jigs and tried several different methods to hold the edges flush. Despite plaster of Paris being a hard material there is plenty of flex in a two-millimetre-thick hemisphere, and I struggled to get the edges to line up. And I was using two-part resin to glue them together, which sets extremely fast. The two plaster-cast halves had an unwavering desire not to stick together accurately, and would distort just as the resin was hardening. If after clamping the halves together I had to make minor adjustments to the spheres, I got resin all over me. It went everywhere and it's horrible to clean off.

Eventually, after many unsatisfactory attempts, I developed a problem-free method: clamp the halves gently, then apply lots of strong (acid-free) masking tape to line up the sides, before tightening the clamp (a repurposed pillar drill). Once the resin had dried, it just required a bit of gentle sanding to remove any glue, and I had a finished sphere.

As I mentioned, had I planned properly with the intention of starting a company from the outset, I would have begun with a smaller sphere and advanced a little faster. However, one of the best things that came out of this initial, steep learning curve was that having battled to master the art of creating a fifty-centimetre sphere, it was possible to graduate to larger globe sizes with relative ease. So, when Ann Marie Peña, the manager of Yinka Shonibare's studio, came calling, looking for five large globes to use in the British-Nigerian artist's sculptures, ostensibly it was not too big a leap. Ann Marie had chanced upon our Kynaston Road shopfront on a night out in Stoke Newington. However, I now encountered an unexpected hurdle. The shop door measured sixty-eight centimetres across, and Yinka's proposed globes would each have a diameter of eighty centimetres. I confirmed the commission (I was hardly going to say no), so I was either going to have to take the glass out of the front window (which I was seriously considering) or find new premises – again.

4

MAKING A MAP OF THE WORLD

Back when everything was made by hand, the maps on globes were produced by engraving the map in reverse onto copper plates, then printing from these on huge mangle-style presses (intaglio printing). I am always amazed when I see these intricately detailed, back-to-front plates, and was lucky enough to see the labour-intensive printing process on some of Coronelli's original plates which are housed in the Louvre's Parisian workshops. The engravings were made by skilled craftsmen doing this day in day out, so perhaps they simply saw the world backwards. Still, they are impressive. More impressive is the detail that cartographers and globemakers managed to include on their creations. It must have been difficult enough to map a small country like the UK, for example, maybe with input from local

'OH, TO BE A FLY ON THE WALL IN CORONELLI'S STUDIO'

cartographers around the country, but how did they manage if they wanted to map continents or the whole world? Naval ships would return from their voyages with the latest mapping, often burning the previous editions, but what about globemakers without these connections? Did they attempt the whole project by themselves?

Maybe they licensed maps from local cartographers in every territory? It must have taken them years. Not only that, but many old globes are quite detailed, with continents and especially their coastlines shown not altogether inaccurately. The interiors of land masses are more vague, with rivers often following generic snake wiggles into the hinterlands, but oh, to be a fly on the wall in Coronelli's studio and to find out his secrets and compromises.

This Bellerby & Co. globe is based on an atlas
made by the cartographer, Samuel Dunn, 1787.

THE GLOBES OF THE SUN KING, LOUIS XIV,
BY VINCENZO CORONELLI, C.1688–1704

Born near Venice, Vincenzo Maria Coronelli (1650–1718) was a Franciscan monk, publisher, encyclopedist, cosmographer and cartographer, known in particular for his atlases and globes.

Coronelli learned the art of globemaking as a young man, possibly when he was studying for holy orders as a novice with the Conventual Friars in Venice. The Duke of Parma commissioned Coronelli to make him a pair of terrestrial and celestial globes, which were so magnificent that the French ambassador to Rome, Cardinal Duke of Estrées, invited Coronelli to Paris in 1681 to make a set of globes for the King of France, Louis XIV.

Globes worthy of the 'Sun King' had to be more impressive than those made for a mere duke, and so Coronelli and his team of craftsmen set about making two matching globes of monumental size. Coronelli formed the spheres using 120 spindles of bent timber about 3 metres (10 feet) long and 10 centimetres (4 inches) broad at the equator. He coated the wood strips with a thick layer of plaster and covered this with a strong, coarse fabric. The globes were then wrapped in two more layers of a very fine fabric which provided the canvas for the highly detailed maps and figures.

Coronelli spent two years in Paris directing the work on the globes before returning to Italy, where he set up a globemaking studio in Venice. It is thought that the globes were finally completed some twenty years later; at least, Louis didn't take possession of them until 1704. No wonder there was a lengthy time frame, because they were a veritable feat of construction, exceptional in their dimensions, each weighing over 1,800 kilos (2 tonnes) with a vast diameter of 395 centimetres (13 feet). The finished terrestrial globe boasted the most advanced cartographic knowledge of the day, including the latest information from French explorations in North America by explorer and fur trader René-Robert Cavelier. But it is much more than a simple spherical map of the world; Coronelli's globe reads like an encyclopedia, covered with calligraphic texts, and large cartouches on an array of subjects, from pearl fishing to the cinnamon of Ceylon.

King Louis XIV installed the globes in separate specially constructed pavilions at his recently completed pleasure palace of Marly-le-Roi, seven kilometres north-west of Versailles. Each pavilion came equipped with books, maps, desks and chairs, all carefully designed to encourage contemplation of Coronelli's glorious masterpieces.

Before his death in 1715, Louis XIV entrusted the globes to the royal library. They were moved from Marly to the library in 1722 and thus escaped destruction during the French Revolution. Today the globes are in the collection of the Bibliothèque Nationale in Paris.

In the modern world, by comparison, information is everywhere and accessible. While I had been focusing my attention on the construction of the globe, for the map that would cover it I planned to rely on the latest technology. At first glance, it seemed a relatively simple task to find a suitable digital world map and license it. I would then simply run the transformation software program on the map, and, hey presto, it would instantly become twelve gores. I could then focus on learning how to apply these to a sphere. What I did not anticipate was the failure of many mapmakers to update their products (just check out commercial globes and maps at your local store), nor the hours it would take to check out all the data, nor the fact that at the outset I did not know what type of map file I needed, what projection I needed, what software I needed, nor how I would actually print the map. Oh, and what paper to use with what ink? So many questions.

I was limited in my choice of map, having established early on that I required a digital vector file. I had no idea what that was either, but I soon learned. Digital photographic images are made up of pixels; as you enlarge the file, the pixels become visible and the image loses clarity, or pixelates – as you often see on low-resolution digital photos. A digital vector file, on the other hand, is a computer graphic that uses mathematical formulae to render its image, so that the solid lines of colour and text can be increased or decreased to any size while always maintaining full definition. They don't pixelate. I also found out that I needed an equirectangular projection – one that when warped into gores accurately represents the shapes of continents and oceans.

To work with vector files, I needed to install Adobe Illustrator software. Notwithstanding its numerous positive attributes and

what you can achieve once skilled in the application, and with the greatest of respect to its creators, this must be one of the most counterintuitive computer programs ever written, or certainly that I have had reason to learn. Its use is governed by learning rules and functions rather than intuition, and there are often many ways to perform the same task. It rewards learning and repetition, like something from my school days. Perhaps it was written by a teacher?

photograph taken from his cockpit window to verify the Aral's new shoreline.) There were also missing countries (especially in the Pacific), places spelled or capitalised incorrectly, labelled with disused names or just omitted, especially in lesser-known areas like Tibet and Mongolia. Some place names even seemed to have been made up. I understand that once you've printed something you can't just throw it away when a new country appears, but this was a digital map, and the company just hadn't bothered to update it.

'I UNDERSTAND THAT ONCE YOU'VE PRINTED SOMETHING YOU CAN'T JUST THROW IT AWAY WHEN A NEW COUNTRY APPEARS'

Not many map companies offered digital vector formats, I discovered, so I had limited options. After weeks of searching, it seemed there were few companies offering maps that met my requirements. However, I found one with a suitable world map and duly paid to license it for my own commercial use. I downloaded the map, and it seemed adequate for my requirements, but over the coming weeks as I began to edit it, I realised that it contained hundreds of errors. By this stage, Google Maps was online, so I had a quick and relatively reliable reference to check against, as well as my trusty Times Atlas of the World. There were numerous towns marked in Greenland, for instance, that when checked using satellite view on Google Maps turned out to be little more than settlements of half a dozen tin huts. Dar es Salaam was marked as the Tanzanian capital despite the 1996 change to Dodoma. The Aral Sea still featured in all its full glory, despite having long fragmented into separate lakes. (I later made a globe for a pilot who sent me a

All I had to do was master Adobe Illustrator, then start editing. I then discovered that all the lines on the map were not joined up. Anything other than a minor river was a series of unconnected lines, so that all coastlines, for example, were made up of thousands of these small lines. I didn't know the consequence of this until later. I was trying to edit the map on my laptop, and given that the program was a huge file, each time I shifted the map across the fifteen-inch screen to work on a fresh piece of text, or to zoom in and out, the laptop struggled to keep up. I spent long days editing into the early hours, much of the time sitting watching the image lurch across the screen as I waited for my computer to catch up with me. If I did one operation incorrectly, which was almost impossible to avoid, I then had to undo the task, save the file (five-minute delay) and reopen it (five-minute delay) before I could advance.

THE OLDEST KNOWN MAP OF THE WORLD

Mapmaking pre-dates virtually all other forms of written communication. The prehistoric paintings found in Lascaux caves in south-western France show depictions of animals and human figures, and abstract signs, some of which are now thought to be Palaeolithic star maps. Primitive maps such as these are proof of our ancestors' first attempts to chart their local surroundings and, more ambitiously, to understand the world beyond their limited geographical experience.

The earliest surviving maps are believed to be those engraved on clay tablets in ancient Babylon. Archaeologists discovered one clay tablet fragment, thought to date from around the sixth century BCE, on the banks of the Euphrates River in the late 1800s in Sippar, Iraq. Known as the Babylonian Map of the World, or Imago Mundi, it is generally accepted as the earliest map of the known world. This remarkable artefact, now in the Babylonian Room of London's British Museum, measures 12.2 x 8.2 centimetres (4.8 x 3.2 inches), about the size and shape of a brick, and shows the world as a disc, surrounded by a ring of ocean, labelled as 'bitter river'. At the centre is the Euphrates River and to the north, the city of Babylon.

The map also shows a channel and swamp at the mouth of the Euphrates, a mountain, as well as unknown outer regions of the world, represented as an eight-pointed star shape reaching beyond the ocean. In the north-eastern 'region' is the inscription 'Where Shamash [the sun] is not seen', reflecting the fact that the sun rises in the east, crosses the heavens, sets in the west, and then returns to the east by, as the Babylonians believed, passing through the underworld.

Who carved this map into clay, and why, is a mystery. It was surely not intended for navigation or exploration; rather it is thought to be a mythological vision of the world and its creation – at least, as much of it as was imagined by the ancient Babylonians.

MADAGASCAR

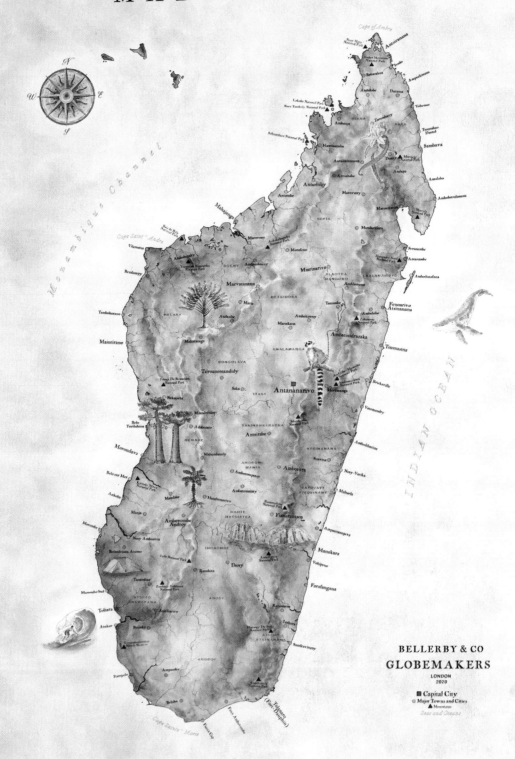

BELLERBY & CO
GLOBEMAKERS
LONDON
2020

■ Capital City
◉ Major Towns and Cities
▲ Mountain
Seas and Oceans

LEARNING CARTOGRAPHY
(OR NOT)

Like many contemporary maps and globes, the map I had licensed had a very simple layout. The format was functional, but felt basic and lacked clarity, its only redeeming feature being the occasional use of alternative, different-sized or bolder fonts. Overall, however, the text was just slapped down regardless of any other elements on the map. With almost no exception all the words were placed horizontally. Consequently, across the map there were numerous places where two or more pieces of text had collided, or lettering overlapped a straight horizontal coastline or other cartography. The result was illegible text.

Having never studied cartography, I didn't know the rules, but given that I was making this globe for my father, I didn't feel bound to follow any established etiquette, especially when I didn't like what I saw. I therefore decided to start again pretty much from scratch. I stripped the map back to the coastlines and repositioned almost every single town name, often carefully placing each individual letter to give the map a more irregular and handmade texture – legible but also aesthetically pleasing. This was such a long process that, despite my care, the globes I made during the first few years still had idiosyncrasies.

Today, we have two full-time cartographers who update maps as climate change and political events alter the size of lakes and the borders of countries, using Google Earth as one of several resources. For example, since I made my first globes, the Aral Sea has diminished further; and when in 2018 the King of Swaziland changed the name of his country to the Kingdom of Eswatini, we immediately amended our map, though we kept the old name in brackets for several years. The world is always changing. A globe is and has always been a record of a moment of history, which is why we often include historical information like the names of leaders and demographic and economic facts, which can only ever be fully accurate at the instant we print the map.

FINDING A FONT

Over the months, as my map editing dragged on and the trials with the plaster of Paris spheres continued at Kynaston Road, I would create a little installation in the shop window showing processes and works in progress each evening before I went home. Often people would stop by the following day and comment. It wasn't entirely clear what I was doing (I didn't have a product), but as I hadn't yet perfected my techniques, I found it both encouraging and also a bit strange to have people observe my trials and errors. However, most people were just passing the time of day, and while it was interesting to chat, there came a point where I realised that some of them had too much time on their hands. I could lose hours.

He introduced himself as James Mosley and suggested very diplomatically that the font I was using on the map was, if not incorrect, then perhaps not suitable. He said that he would go away and come back with suggestions in the morning. By now other passers-by, on discovering what I was trying to do, were sharing their ideas on how to make globes, their opinions about the designs and colours I should use, and how and where to market the finished products; indeed, there seemed to be so much globemaking knowledge in Stoke Newington, I was surprised that a collective hadn't formed to found a rival company.

'BY NOW OTHER PASSERS-BY, ON DISCOVERING WHAT I WAS TRYING TO DO, WERE SHARING THEIR IDEAS ON HOW TO MAKE GLOBES'

One day I noticed an elderly gentleman staring intently at the work on display in the window. By this stage I had discovered it was easier to leave the door locked so I could decide whether I wanted to engage. The gentleman seemed relatively normal, a bit scruffy, kind of like a science professor, so I opened the door and said hello. My window display consisted of strings of practice paper gores I had been working on the previous day. I still didn't have a finished morphed map so my gore sections cut off swathes of detail, but he didn't pick me up on that; he seemed more interested in something else.

I didn't expect to see James again, but the next day he returned with some samples of the font he suggested I use. I took his offer of advice politely and bade him farewell. Given that the font on a globe is one of the most important elements in its design, I wasn't necessarily going to take the suggestion of a random man, though oddly he didn't hang around. However, perhaps the world's leading authority on fonts and the man who had run the St Bride Foundation typographical library for forty years was someone that I should listen to. After he'd left the shop, I'd looked him up on Google and discovered his Wikipedia page. Amazingly, this fount of typographical knowledge lived only a few doors away. James came back a few days later and kindly lent me his own unique handmade font, Britannia, which I use on many of our globes to this day. I willingly traded the use of this beautiful font for a globe, though he was happy just to help and asked for nothing.

Around this time I also decided that, just like any firm, we needed a logo, so I had to think about a cartouche for my new globe.

I came up with a simple design, which I used for around a year or so. And then, not long after my happy encounter with Mr Mosley, another neighbour came into the shop. James Hislop was and is the resident science curator at Christie's, and thus their globe expert, and he was thrilled to find out what I was doing. We had a long chat about the industry and how I should market the company. He predicted I would have customers queuing around the block. Maybe my ducks were starting to line up.

CARTOUCHE DESIGN

According to map historian Edward Lynam, simple strapwork cartouches framing the titles of maps first appeared in Italy in the sixteenth century. Over the centuries, cartouche design evolved and varied according to cartographer and period style. By the seventeenth century, engravers were often adding extremely ornate and intricate architectural and figurative elements to their cartouches, including coats of arms, mythological figures and beasts. In Europe the ornamental cartographic cartouche reached its apogee in the baroque period, during the late-seventeenth and early-eighteenth centuries (see the Coronelli globes for Louis XIV, for example), with designs becoming more fluid, incorporating the human body, exotic animals and plants. By the end of the eighteenth century, ornamentation in cartography had fallen out of fashion, and a more sober style developed with a preference for simple oval or rectangular frames containing scientific and cartographic inscriptions.

MORPHING

Back in 2008–9 there was very little software available to morph a wall map into gores, so while I was busy editing, my friend Jon was writing the code from scratch. Inexplicably his full-time employer then sprang news of a relocation on him. Rather than working from home in leafy Berkshire, he was going to an office in Karachi, which was not ideal for either of us. It would be fine, he told me; he could still work on the code for morphing the map. However, he soon found out that his workload had also morphed, taking up much more of his time. He was clever and had usually been able to do his day's work before lunch, but now he had no spare hours during the day, and when he said goodbye to his submachine-gun-wielding bodyguard in the evening, he understandably needed to relax rather than spend the night in front of a computer monitor.

Nevertheless, he persevered. The code he wrote required me to save the vector map within his program and run a task before reopening it, which would then shut down my computer for about half an hour each time, rendering it lifeless. It was difficult to tell if anything was happening at all. Remarkably, it worked the first time, producing twelve perfectly shaped gores. In the end, he delivered the program around the end of 2009, about a year behind schedule.

However, because the lines on the original map files did not join up and because the morphing program effectively sliced the map into sections, there were numerous problems with the text and map lines across the gores. It quickly became apparent that I needed to edit each individual gore section, and not the original full rectangular map projection. All the editing work I had done on the map file had therefore been in vain. I had to do it all over again on the new morphed version of the artwork. Months more work, although at least by now I had learned the program well enough, so I had more confidence and could work much faster.

'PUBLISHING' A GLOBE

Like an author whose book is 'published' by his publisher, the cartographer, who publishes an original map or map gores, and the globemaker, who also 'publish' their work. This is still the case today and makes sense since the globe is simply a printed 3D version of an atlas. So, if you were to simply steal someone's work by reproducing an old globe, it would count, technically, as plagiarism.

However, in the sixteenth century when, like authors, early cartographers such as Gemma Frisius and Gerard Mercator were publishing their globes and maps, the copyright laws were less about protecting the author or cartographer and more about state rulers maintaining control over what was printed. Thus, a government could censor anything it considered seditious or blasphemous and condemn its author to death for heresy.

THE POLITICS OF GLOBEMAKING

When I started editing, I didn't really give a thought to which countries I would include on my globe and which not. I was going to include them all, obviously. But it's quite easy to forget that there are numerous border disputes, and, as I discovered quite quickly, what rule to follow was not obvious at all. Indeed, it only takes a minute to realise that globemaking is a political minefield. Some countries are highly sensitive about their borders. In India I could go to prison for selling a globe that shows its border with Pakistan following the line that the government of Pakistan believes is correct. In Lebanon a globe with Israel marked on it will not get through customs. China will not accept any map or globe that depicts Taiwan as an independent country. Chinese customs impound and destroy such items, or simply return them to the sender after months of silence. Even raising the subject is forbidden; this book cannot be printed or sold in China.

Many countries have national, often nationalistic, cartographic societies and institutions. One might think the United Nations would promote an internationally agreed cartographic representation of the Earth, but that body is so careful not to offend the world's biggest wallets that it fails to recognise actual countries. The reality is that there is no international cartographical standard. Every country sees the world differently based on its own history but also to a greater or lesser extent from a nationalistic and strategic point of view, and this is reflected in the cartographical information it accepts.

I felt that the best way forward would be to represent each country as its population, democratically elected politicians, monarchy (whether popular or absolute) or dictator saw fit. If I sold a globe to an overseas client, I had to avoid offending the receiver (or their customs personnel), though in practice this means we have to edit our maps depending on where the globe is to be shipped, so we have versions for China, India, Lebanon, etc. Many countries don't care a jot, but many do, and some have good reasons for caring. We mark disputed borders as disputed. We cannot change or rewrite history.

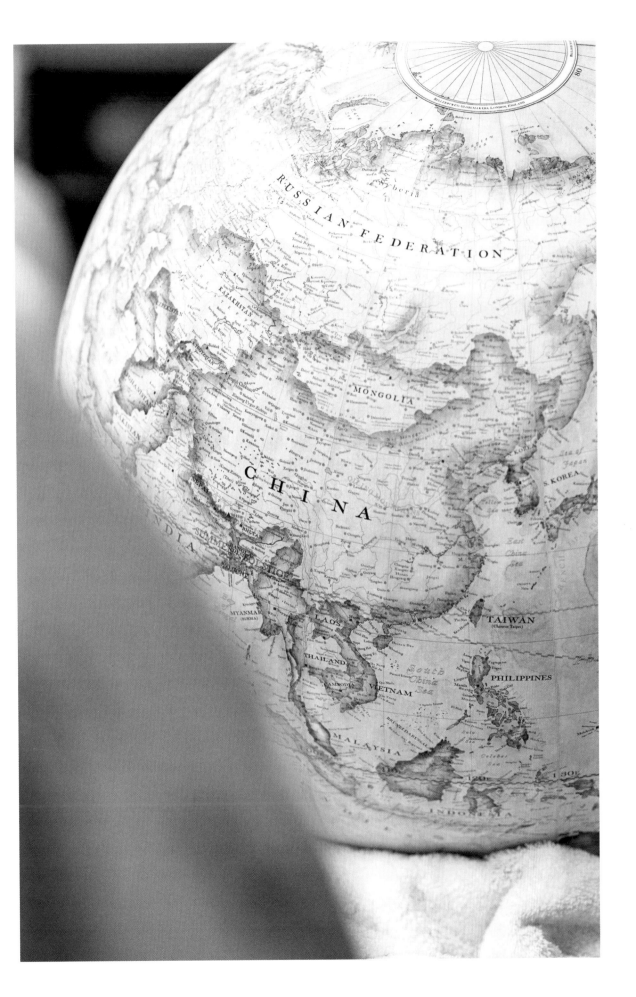

PAPER AND PRINTING

One of the biggest questions I had at the outset was which paper to use on my globe. Of similar importance was the printing process and which inks would stand up to being drenched and stretched. My limited knowledge of the printing business led me to believe that the only option was lithographic printing, an old-school process where the paper runs four times through the presses, since this uses solvent-based inks which supposedly last much longer. The map editing process was slow and ongoing, and I had months ahead of me before I would need to start printing the finished map sections, so I had ample time to go traipsing around out-of-town business parks in my search for a suitable printing firm. I was also fortunate that there had quite recently been significant industry advances.

wide-format technology was breathtaking, far sharper than a lithographic process could ever achieve. I found a local commercial print shop; I just needed to source my paper.

I had read a few articles about there being a specific paper for globemaking, but I had no idea where to find it or why it was so special. Michael, the owner of the local print shop, had decades of printing experience. He almost fell out of his chair at the suggestion that anyone had ever produced paper exclusively for globes, never mind that it might be an option today. Early globemakers would have used handmade papers by default (the only paper available before the Industrial Revolution), which are more loosely woven and so much easier to manipulate. But even so, in his view – one that I couldn't argue with – the idea that someone was making special paper for globes

‘SIR, I WOULDN’T RECOMMEND ANY OF OUR PAPER MEDIUMS FOR THAT’

Months of research, endless conversations and trying to come to terms with the rigid requirements of litho printing (no edits and huge volumes) were all rewarded when I learned that over several years there had been a transition from solvent-based inks to more ecologically sensitive products. Affordable wide-format printing, which you see in high-street print shops, had also evolved and thus entered the equation. The new eco-inks had guaranteed lifespans of up to 200 years. Not only that but the print quality of the new

seemed fanciful, if not utterly ludicrous.

I called up Canon customer services and spoke to one of their agents. ‘My plan is to print an image on paper, dunk it in water, then stretch it over a sphere,’ I said. ‘Which paper and which ink can you recommend?’ ‘Sir, I wouldn’t recommend any of our paper mediums for that,’ came the reply. Evidently, no one had posed this question before. They didn’t have a clue, because they didn’t need to. In fact, no one I spoke to, not even Michael, had the slightest idea which paper I should use. Given that the printed gores would then be hand-painted, my chosen paper also had to take watercolours well, so it needed to be absorbent. Michael and I discussed the options, and he gave me samples of several dozen different grades to see how they worked.

The next make-or-break moment would be testing the new eco-inks. Given that these were not solvent-based there was an obvious question mark over their ability to hold on the paper when I dunked it in a bowl of water. Every paper was likely to absorb the inks differently, and I just didn't know how permanent they would be. If the ink ran or faded when the map was wetted, it would derail the entire project. Amazingly, when I soaked one section of map in water immediately after printing, I was delighted to discover the eco-inks were clearly permanent.

Since 2012, we have been using our own wide-format printers in the studio, allowing our maps to be as up-to-the-minute as possible and giving us complete flexibility in including customers' personal requests.

CARTOGRAPHY AND PLAGIARISM

Back when mapmaking was still a new profession in America, some cartographers would include invented towns on their maps. This was not to confuse travellers, but a smart ruse to identify copycats. Stealing and copying another cartographer's map was common. But if a non-existent town was spotted on a competitor's map, it was easy to prove forgery.

The first of these bogus towns on a map of North America was Agloe, New York, in an edition by the General Drafting Co. in the 1930s. Agloe then resurfaced on maps produced by Rand McNally, when cartographers for the company found that someone had opened the Agloe General Store at the exact spot of the fictitious town, thus making the location 'real'.

Agloe is not the only copyright trap that took on a life of its own. Take Argleton, a town in north-west England – or, more accurately, an empty field in West Lancashire. For a while, Argleton appeared on Google Maps, on various websites and in listings for estate agents and employment agencies. You could also search for the weather forecast in Argleton or for the name and number of a local plumber. The businesses were real, but they existed in other nearby towns in the same postcode area. In 2009, Google erased Argleton from its digital map, putting the mistake down to 'occasional errors' in its mapping information.

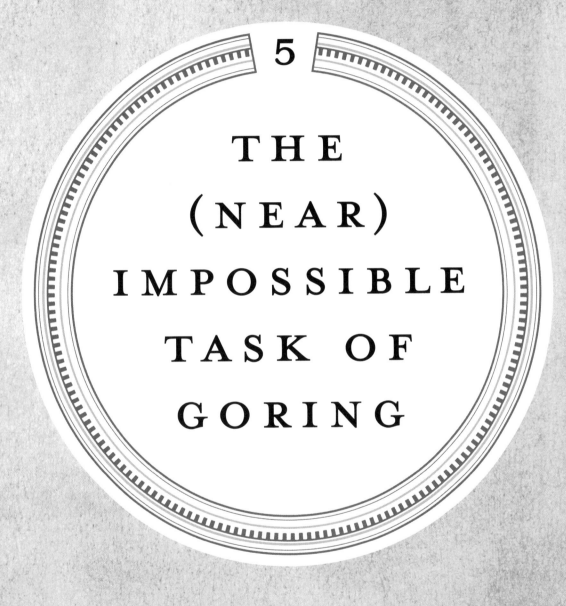

5

THE (NEAR) IMPOSSIBLE TASK OF GORING

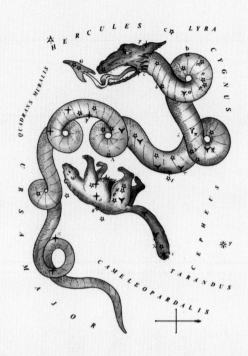

M odern globes are valued above all as objects of beauty, cherished for the incredible craftsmanship and mystery of their construction and the artistry of their map designs. This makes them extremely costly and as such they are formidable status symbols. But historically they represented much more than this. During the Age of Exploration, the period of European overseas expansion from the early fifteenth to the eighteenth century, the demand for maps and sea charts was at a peak. Globes were practical and efficient scientific and

'DURING THE AGE OF EXPLORATION THE DEMAND FOR MAPS AND SEA CHARTS WAS AT A PEAK'

educational tools, which allowed scientists and scholars to study and explore the relationship between the Earth and the celestial bodies. A terrestrial globe was an invaluable asset for merchants and businessmen when plotting new potential trading routes overseas. And for sailors navigating across the oceans, using only the positions of the stars as their guide, a celestial globe, often used alongside its terrestrial pair, was a crucially important practical tool. Until 2010, the Admiralty installed a celestial globe in every Royal Navy ship, just in case.

Today, thanks to satellite navigation, the problem of reliably plotting the position of a ship at sea has been solved to within a few centimetres of accuracy. But before the invention of the magnetic compass, the search for a reliable method of calculating direction, distance and position (latitude and longitude) at sea took millennia of study by astronomers, cartographers and navigators, and involved many of history's most celebrated scientific minds.

It was Eratosthenes, in the third century BCE, who, having calculated the Earth's circumference, first proposed a grid system of lines for a map of the world. Knowing that the Earth was a sphere, and knowing its circumference, the location of the equator could be mathematically calculated. Distances north or south of the equator, the lines of latitude, could therefore be calculated from the relative position of a celestial body – the sun or Pole Star, for example. Calculating longitude, distances east to west, was more problematic, however, as there is no tangible, natural reference point from which to measure; it's a question of time.

A century after Eratosthenes, the Greek astronomer, geographer and mathematician Hipparchus devised a system of coordinates to identify the location of places on Earth. His lines of longitude were based on dividing the globe into 360 degrees, so that longitudes could be described as degrees east or west of the prime meridian, zero (in his case, Rhodes), just as we do today. Hipparchus proposed a method of determining longitude by comparing the local time of a lunar eclipse at two different locations. Every twenty-four hours the Earth revolves through 360 degrees, so one hour of time difference equates to 15 degrees of longitude ($360 \div 24 = 15$). The method was not very accurate, due to the constraints of the available clocks, but his idea was sound. Longitude could be established by accurate knowledge of the time.

PTOLEMY'S INSTRUCTIONS ON HOW TO MAKE A MAP OF THE WORLD,
C.150 CE

Claudius Ptolemaeus, known as Ptolemy (c.100–c.170 CE) was a Greek mathematician, music theorist, astronomer and geographer working at the royal library in Alexandria. Today this polymath is renowned as one of history's great cartographers.

Ptolemy was the author of several books including the *Geographical Exposition*, or *Geography*, a guide to drawing maps in which he summarised the cartographical and geographical knowledge of his Greek predecessors. Although people had been making maps constructed on mathematical and scientific principles since the time of Eratosthenes, Ptolemy improved on previous attempts at mapping the world by using a rectangular grid of longitude and latitude lines on which he could fix the position of a place on the Earth, based on the reports of estimated distances and directions from returning travellers and seafarers. *Geography* included an extensive table of the coordinates of approximately 8,000 places which could be plotted on these lines to create a map. His great innovation, however, was to realise that, for this map to be accurate, his grid needed to account for the curvature of the Earth over large areas of sea or land. To represent the three-dimensional globe on a flat map, therefore, Ptolemy used projections with curved lines, which reduced the distortion.

His system for 'spherical coordinates' was valid and a considerable step forward in cartography, and thus the accuracy of navigation. However, his world map was not perfect. This was because Ptolemy had used a much smaller estimate of the circumference of the Earth than Eratosthenes' more accurate one. *Geography* was translated into Arabic in the ninth century CE, shaping the geographical knowledge and cartographic practices of mediaeval Arab geographers, while Western scholars were ignorant of Ptolemy's great work for more than 1,200 years, until it was finally translated into Latin early in the fifteenth century. Thereafter its influence on Renaissance, pre-Columbian cartography was profound and long-lasting, if not entirely positive. Due to Ptolemy's error in using the smaller estimate of the Earth's circumference, adventurers like Christopher Columbus perhaps made over-optimistic assessments of the viability of circumnavigating the globe. Nevertheless, because of his findings and detailed instructions on mapmaking in the *Geography*, Ptolemy is often hailed as the father of geography.

The measurement of longitude is of vital importance to both cartography and navigation by helping to establish points and distances on the Earth, not least to facilitate safe ocean travel. For centuries the problem was that to calculate longitude a ship's captain or traveller needed to know the time both where they were and the time at a fixed reference point – a known location. Without a reliable clock to show both times simultaneously, the conundrum remained. In the eighteenth century, an era of colonial conquest and the burgeoning of maritime trade, to guard against the loss of life and perhaps more importantly profits, the safety of sea travel was of pressing concern. As competing nations vied to rule the waves, such was the need for an accurate means of calculating longitude at sea that the British government passed the 1714 Longitude Act, creating a financial reward for the solution to the age-old enigma of finding a ship's precise location on a voyage.

The man remembered for solving the problem of longitude is John Harrison, a self-schooled clockmaker from Yorkshire. His travail to build an accurate and fail-safe timekeeper that could withstand the rigours of a lengthy sea voyage was a lifetime obsession. Finally, in 1766, forty years after his first prototype was tested, Parliament rewarded John Harrison for his services to the nation – though they did not pay all he felt he was rightly owed.

In winter 2011 I was firmly established in my shop on Kynaston Road. It probably looked odd to passers-by; I had opened a shop but had nothing to sell. Given that I was now a fledgling business, though, I set myself production targets. Each new week I set aside time to make eight to ten hemispheres, which with collateral damage translated to about four finished spheres. Then, after preparing them, I continued to learn the process known as goring: wrapping a flat piece of very thin paper over a sphere, without flaws. Gores are surfboard-shaped pieces of paper that fit on a globe, thicker at the equator and tapering to a point at each pole. They can be made up of one, two or more pieces, divided lengthwise.

A globemaker's most important study is to understand the feel, elasticity and weave of the paper so it neither tears nor turns into papier mâché from being wet and manipulated. This requires patience, an understanding of how the paper, glue and water interact, and most importantly learning how to move your body very, very slowly. It also requires accuracy. You are dealing with tiny margins. If you are loose placing just one section of the map, by the time you reach the last gore, you might be overlapping enough to swallow a small ocean. You therefore have to understand the simple mathematics of the challenge, spot any error before it is too late to rectify and constantly re-measure.

THE FIRST PRINTED PAPER GORES:
MARTIN WALDSEEMÜLLER, 1507

During the sixteenth century in Christian Europe, as new trade routes were discovered and new geographical information was pouring in from around the world, ambitious kings and queens, hungry for colonial conquest, as well as sailors, explorers and merchants, all needed up-to-date, accurate maps and sea charts.

At the same time several factors converged to make this possible. Johannes Gutenberg's groundbreaking innovation of printing with movable type in the 1450s allowed a wider public to access all kinds of texts and knowledge. At the end of the fifteenth century and the beginning of the sixteenth, Latin translations of Ptolemy's seminal works, *Geographia* and *Almagest* (astronomical text), were printed for the first time, making them more available.

Newly apprised of Ptolemy's cartographical instructions, and as their scientific understanding of the Earth and its place among the planets grew, Renaissance mapmakers were inspired to devise new ways of representing the fast-evolving view of the Earth and heavens. They too were using the new printing presses to disseminate the fruits of their labours, making them more accessible to the wider public.

The German cartographer Martin Waldseemüller (c.1470–1520), working in Saint-Dié, France, is credited with being the first person to print paper globe gores from woodcuts in 1507, although no globe using them survives. He was also the first person to create a printed wall map of Europe, and some consider a set of his maps printed as part of Ptolemy's *Geography* in 1513 to be the first modern atlas. With his collaborator Matthias Ringmann on their 1507 map Universalis Cosmographia, Waldseemüller is also notable as the person to give the name 'America' – after the Italian explorer Amerigo Vespucci – to a part of the newly discovered continent, which he said was 'proper to call a New World'.

The printed world globe, as we recognise it today, appeared in the early sixteenth century, and for the next couple of hundred years, there were some incredible cartographers and globemakers who crafted the most beautiful depictions of the Earth and the celestial spheres. As knowledge of the Earth's geography and of astronomy increased, these became more accurate.

At the beginning, as I was learning the process, I started with gores split into two at the equator. Today, with more hands available, we tend to use one piece north to south, but this is difficult, especially on a large globe when the paper can be over two metres long. Historically, makers have used up to four segments, usually cut at the tropics and equator. Joining the gores at these points was the most difficult skill to learn; it required hundreds of attempts to get it right, and also learning the limits of the paper to understand its properties fully. It had to be pushed to its elastic limit and further to ensure I knew its breaking point.

their secrets (which they probably wouldn't anyway), there were none that I wanted to learn from.

The first process I needed to work out was how to mark the globe so that I could accurately place the gores. I could establish a North Pole without much difficulty, but trying to plot the corresponding South Pole on the same sphere was not as easy as it might seem, without sophisticated imaging machines. It required a jig and repetitive marking to confirm the measurement.

I then fashioned a giant compass out of two strips of wood with a nut and bolt for a hinge, a sharpened nail protruding from

'THE FIRST PROCESS I NEEDED TO WORK OUT WAS HOW TO MARK THE GLOBE SO THAT I COULD ACCURATELY PLACE THE GORES'

Once I had started making globes, I seemed to notice them everywhere. I also noticed how badly made many of them were, particularly those manufactured since the Second World War; latitude lines that did not meet across gores seemed to be a recurring theme. Why weren't globemakers lining these up? Some makers, to prevent the paper ridging at its edges, cut little triangles out of the map and thus lost text. As I studied more globes, I even found makers from the 1960s and 1970s who had overlapped their gores to the extent that they had wiped out entire land masses, Alaska and Greenland being frequent casualties. Why were they doing their job so badly? There seemed little point to me in spending time researching a project only to produce a poor-quality item. I concluded that even if I had been able to find an existing globemaker to teach me

the end of one strip of wood and a hole to hold a pencil at the end of the other. I also made a wooden bung to fit in the hole that had been drilled at each pole and then fitted the nail on the giant compass into the bung so that I could use the compass to mark concentric rings on the globe as latitudinal guidelines for the gores. It required a steady hand and working slowly. I was using a pencil to mark a very abrasive surface, and the angle of the pencil on the sphere tended to open the compass, so if I rushed, my line tended to both bleed and increase and I lost accuracy. I first did this from each pole to locate an accurate position for the equator, which obviously needed to be equidistant from them. Then I added a series of concentric circles at ten-degree intervals so I could line up the horizontal latitudes on the map.

There were only so many blank pieces of paper I could feel passionate about; I wanted to see how mapping on a sphere looked. With Jon in Pakistan and his program to morph the map still seemingly months away, I had started to work with outline gore shapes laid across a standard map projection so at least I could get an idea of what a map looks like on a sphere. The outline gores needed to have latitude lines in place, since I could see that keeping these aligned was the obvious way to line up the map.

Using a scalpel blade, I cut the gore sections along the edge of the longitude lines to within two-to-three tenths of a millimetre. It was far easier to get an accurate cut with a scalpel, although each blade only lasted around two metres of cutting. The cut had to be a single smooth line; though this involves practising around 50,000 times. (It brings perspective to the gouge that a surgeon put in my stomach when I had a kidney stone removed in my early twenties.)

I had limited experience of working with paper – I had obviously used the dry material in everyday life and in art classes and I'd played with making papier mâché objects as a child – but it seemed to me that to fit each gore to the solid sphere, I would need to soak the paper to relax the fibres, but bear in mind that it would disintegrate if I left it in water for too long or just fall apart if stretched too far. I knew nothing about paper making other than that none was specifically designed for this job. As I worked, though, I was pleasantly surprised to find that the wet paper held up remarkably well under tension.

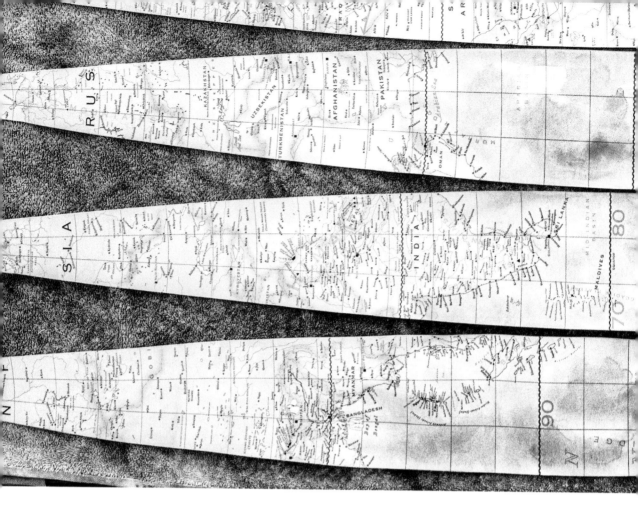

It was going to be a process of learning through trial and error, but there was a recurring issue. I planned to test many different papers and was looking for a paper that stretched just enough. Perhaps because at this stage I was using an external printer, so not seeing how the printing was physically carried out, it wasn't until I had at all, and some stretch before falling apart, while others lose their top layer under the stress of stretching. Not only that, but different batches of the same paper from the same manufacturer are not identical. But what I couldn't work out was how some days I was able to stretch the same paper sufficiently and other days it just

'I NEEDED SOME INSPIRATION, SO I WENT TO VISIT THE NATIONAL MARITIME MUSEUM'

done many trials over many months that I had a eureka moment – one that led me to one of the secrets of modern globemaking.

Yes, paper stretches when it's wet, but not identically. Some papers barely stretch wouldn't budge. And then I reasoned that since it is made on a roll (we use paper in hundred-metre rolls), it is probably important that paper isn't stretched in the direction of the roll, but across the width

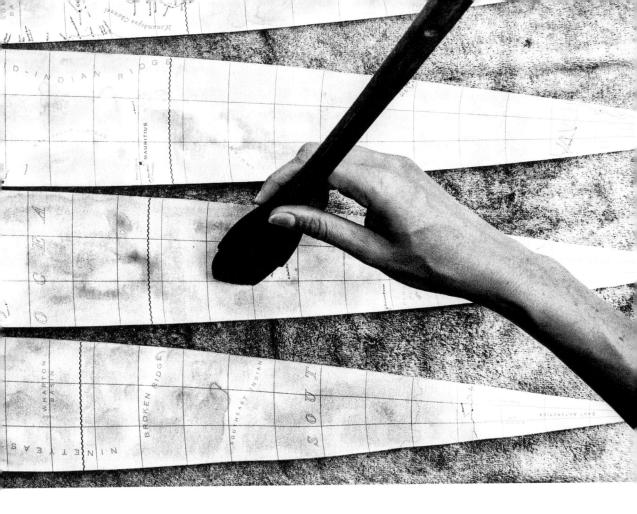

of the roll. A similar principle applies when cutting a dressmaking pattern on the bias to create a naturally stretchy garment. This annoying anomaly meant that the plane on which the printer set up the maps each day had a direct correlation to the success of my day's work. Printed across the width of the paper and I was in the pub at 6 p.m.; printed along its length and I would still be tearing my hair out at midnight. For weeks on end, this led to many fraught moments until I realised what was going on. This was just one of four or five neither planned nor expected breakthroughs that allowed me to take great leaps forward in working out the process. Had they not happened I might not be making globes today.

I needed some inspiration, so I went to visit the National Maritime Museum, which boasts a collection of over 400 globes, both celestial and terrestrial. Making a quick sprint to get ahead of a horde of shrieking schoolchildren bearing down on the entrance, I began my tour. I saw one little globe, then another couple upstairs; eventually I think I found five or six in total. Where were the other 394-plus? I asked to speak to a curator but no one was available. Back home, I found an email address and sent off my enquiry. I learned that, like most museums, only a fraction of their collection was ever on display.

Filling in the institution's overly bureaucratic forms giving the reasons for my proposed visit to their archives (this is not a criticism of the lovely people who work there), I was eventually deemed worthy of access, and several weeks later I made my way to a mysterious location – an unprepossessing warehouse on a trading estate – to see the globes. Entering a huge vault, I walked through rooms of paintings and other priceless artefacts far more interesting than those in the museum, before discovering the globe section, and there, in prime position, on show to cleaners and security and the occasional visitor, was an original seventeenth-century Willem Blaeu terrestrial globe in a glass case. This is one of the finest globes ever made and worth upwards of £500,000 at auction. I also found a pair of eighteenth-century globes which apparently had been handed over in lieu of death duties. I wonder who had valued them. HMRC? And surely it would be better to give or sell them to someone who might restore them, or at least get money back from them, as they were given to HMRC in lieu of tax. I would prefer them to fall apart in view than fall apart hidden.

The curator told me that many of the antique globes are in such poor condition that they can barely be moved, let alone displayed. They are prohibitively expensive to restore or conserve, and so they sit there unseen. Each day another decaying fragment drops on the floor. I imagined returning to the vault in a few thousand years when in place of those beautiful artefacts, there might just be 400 little piles of dust. Sad.

The Blaeu globe, however, was amazing and beautiful. For me, knowing what I wanted to achieve, this visit was perhaps like a young football-mad kid going to Anfield, or an aspiring dancer going backstage at Covent Garden. If Blaeu could make a globe as magnificent as this in 1603, surely today in the early twenty-first century, I could hope to master the art one day soon.

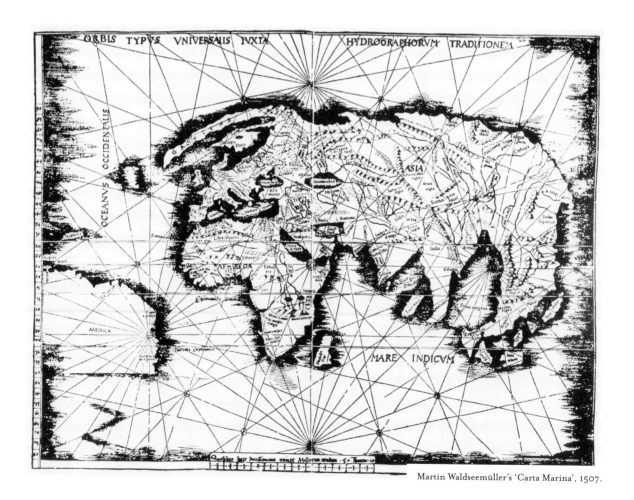

Martin Waldseemüller's 'Carta Marina', 1507.

WILLEM BLAEU'S GLOBES

Willem Blaeu (1571–1638) was one of the seventeenth century's preeminent map- and globemakers. Thought to have been born in Alkmaar in North Holland, Blaeu moved to Amsterdam in his youth, where he worked for his father's herring company. Having developed an interest in mathematics and scientific instruments, he worked briefly as an apprentice to the astronomer Tycho Brahe before becoming a printer and marine cartographer.

Blaeu soon started to publish globes, books and terrestrial maps. He produced his first globe, a small 13.5-centimetre (5.25-inch) celestial model, around 1597–8, and what sets it apart is its innovative illustrative style. Blaeu was a publisher and as such did not engrave his globes and maps himself. For this he employed Jan Pieterszoon Saenredem (1565–1607), who embellished his globes with depictions of figures in contemporary dress and other novel decorative features. Blaeu then published a 34-centimetre (13.75-inch) terrestrial globe in 1599 and its matching celestial globe soon thereafter. Blaeu diversified with various smaller globes, constantly updating them with the latest cartographic information. Then in 1617 he published a matching pair of huge globes, measuring 68 centimetres (26.75 inches) in diameter.

Later with his sons Joan and Cornelius, he continued to publish increasingly ornate terrestrial maps, atlases and globes, whose beauty and lavish adornments rivalled those of his arch commercial rivals, fellow Amsterdam cartographer Johannes Janssonius and the engraver, cartographer and globemaker Jodocus Hondius. The competition was fierce, and as each of these Dutch makers vied to outdo the others, their globes became ever more magnificent and breathtakingly beautiful.

GLUE

The next issue was which glue to use to fix the gores firmly on the naked sphere. This was another important choice where my knowledge was lacking. I felt that I needed to apply the glue to the sphere first rather than the paper but could not allow it to set before I accurately positioned the paper on the sphere. The problem is that if the glue is still wet when you apply a wet gore to it, it simply rinses off with the excess water from the soaked paper. Wait too long and the glue is dry and so it fails to bond. I also had to factor in the difficulty of working with a spherical object.

use for my trials would be wallpaper paste. It was cheap and easily available from the local DIY store. I remembered my father wallpapering but hadn't tried it myself. The paste is thick like semolina and not much fun to work with – very lumpy, gets everywhere and not very strong. Almost every time I stretched a gore over the sphere, thick globules of paste got trapped underneath and punched holes through the paper. It was much too thick, and when I diluted it to get it thin enough it had lost all its strength.

'I THOUGHT THE SIMPLEST THING TO USE FOR MY TRIALS WOULD BE WALLPAPER PASTE'

When I was applying the later gores, the globe would be resting on the surface I had completed earlier with a strong possibility of damage. I also had to make sure that the glue-and-water mix didn't run down and change the absorption properties of the gores I had already laid.

In addition, I realised that I had to use a pH-neutral glue. Many old globes have discoloured due to acidity in the glue, water, varnish and/or paint used in their manufacture, so I wanted to keep all my materials as close to pH neutral as possible. To begin with, though, just to learn the process, I thought the simplest thing to

I figured I would just use ordinary PVA, which at least had the consistency I required, but almost before I started, I had an impromptu visit in the studio from an artist who had seen what I was doing and wanted to work for me. At that stage, I didn't have a job to offer her, alas, but we had a long chat, and it transpired that she was very knowledgeable about working with different glues, so I had another little eureka moment, with almost perfect timing. She told me what I needed and where to get it.

I hadn't found the perfect paper either; I would just have to keep on trying different samples and then, at some stage as my skills improved, I could refine my options. Meanwhile, every day I would try to gore a sphere, and every day I would fail but at the same time make some progress. Some days I would complete half a globe; others, I'd barely manage one gore. Another problem I had was that the glue was so strong (exactly what I needed) that it was far quicker to make a new sphere than try to remove the map I had stuck on the day before, which would take all morning. So, I couldn't just practise on the same sphere over and over.

At the end of each fortnight, I would fill my car with as many spheres as would fit, drive to the local municipal dump and launch them from height into the skips. I tried to do this without drawing attention to myself as I was a little paranoid that someone would try to stop me or fine me for breaking an obscure regulation. At the same time, I wanted to test the tensile strength of the spheres, so I would bring along other junk in the car as a distraction and then chuck them in when no one was looking. Annoyingly the hessian backing was so strong, the launched spheres only ever managed a small unsatisfying crumple, but at least they proved structurally sound.

EXPLORING THE NORTH POLE

The American Frederick Albert Cook, claimed to have reached the North Pole in 1908, though due to a torturous return journey, he didn't arrive back home until well into 1909. He was unable to back up his case with any navigational records and even some of his crew suggested that they had fallen well short of their target. Initially, however, he was somewhat lauded until his former friend Robert Peary lodged his own claim. Peary had been funded by the National Geographic Society, which lent more credibility to his exploits, though they also have been disputed ever since. He was the only member of his crew with navigational experience, and after the only other wayfinder left the expedition, his subsequent records suggested a much faster pace than they had thus far been achieving. National Geographic believed he had been within 5 miles (8 km) of the pole, and more recent attempts to verify his journey have strengthened his assertion.

Both could have got there. Cook, it seems, elaborated an earlier story of having climbed Mt McKinley (now Mt Denali), and later spent years in jail for fraud, so had form for fabrications. Peary's claim equally does not seem to really hold up to scrutiny, including the congressional examination of his daily journal which was surprised by its pristine condition; hardly something that had endured the extremes of a polar expedition.

The first undisputed overland trip to the North Pole didn't occur until 1968, and was led by the American Ralph Plaisted travelling on skidoos, though there had been a number of earlier aircraft flights over the Artic. The following year a more traditional dogsled expedition led by Sir Wally Herbert, a Briton, reached the pole and was the first to arrive there by foot.

The other thing I had to contend with was accurately sizing the gores. The vertical measurement of a gore must be such that when stretched it finds its natural curved length and the edges lose their wrinkles – but no more or they lose the integrity of the curves and gaps appear between gores. The length of each side must almost exactly match the distance from pole to pole or pole to equator, as it barely stretches. Then there is the width; depending on the

If the gore was just a little too long or too short, often it would work for the first six or seven, then go completely awry for the rest. Having started at Alaska, I would get stuck with a panel, usually around Africa, where it would either rip repeatedly or it was impossible to smooth down with ridges on each side, and simultaneously impossible to match the latitude lines with the preceding gore.

At one stage I had a private meltdown,

'THE PROCESS REQUIRED CONSTANTLY MEASURING AND RE-MEASURING EACH GORE TO GET IT RIGHT FIRST TIME'

number of gores used, the width must find a distance of one-twelfth of the globe minus an accurate calculation for the stretch of each gore. Ultimately, if a sphere has 24 gores stretched around it, and if you lay each gore just 0.1 millimetre short widthways, then when you lay the final gore, a gap of 2.4 millimetres has opened. Obviously, a tenth of a millimetre is a tiny adjustment to gauge visually; even half a millimetre is difficult to work to, so minor adjustments as the gore dries are important, as is allowing for final shrinkage when dry. If you overstretch the paper widthways, because there is a lot of tension in the paper and little resistance, when the gore dries it will shrink and gaps will appear. The process required constantly measuring and re-measuring each gore to get it right first time, and ensuring that I allowed enough time between each gore application for the previous one to dry a little – and shrink – but again not too much, so that I stayed on track.

when I was thrown by the printer changing his printer settings over several weeks, so that he was intermittently printing at 99 per cent of scale rather than full size. It got to the point where I had to measure every single gore to ensure it was correct. It was easy to believe that there was something that I just didn't understand, some secret those makers of old had discovered. Maybe the watercolour pigment I was using was degrading the paper? I just didn't know whether it was the pigment, the glue, the temperature, the humidity or me. There were so many variables, it felt like if one of them was wrong I might never find out which. I had to test every possible combination, but there just weren't enough hours in the day.

18 months later, I was many tens of thousands of pounds into my venture and I was still a long way away from being able to do this part of the job. And there was a crucial contradiction I needed to resolve:

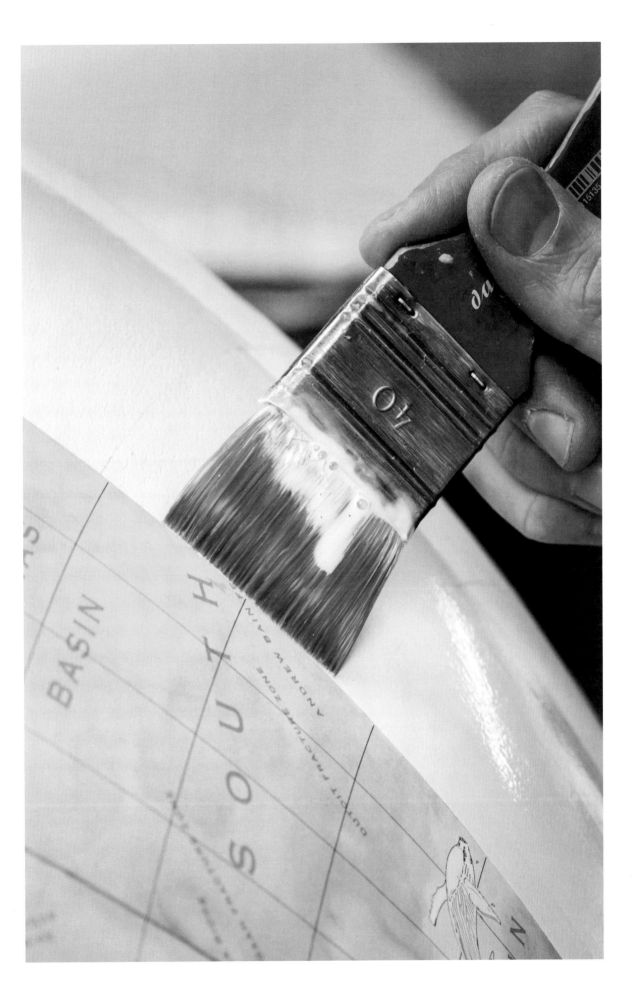

how to apply the minimum amount of pressure so I didn't bruise the paper, while at the same stretching it, which obviously requires exerting pressure on the paper, which obviously results in bruising. Any careless hand movement, even a gentle brush with a fingernail, would damage the fragile paper. Smooth hands are important, I discovered, giving me the perfect excuse not to undertake any heavy manual jobs at home, but I realised the most important thing was to practise mindfulness, to focus on moving my hands very slowly and with consideration. I took up yoga. (Nowadays, the whole studio does a yoga class together every week.)

I just kept on repeating the process over and over again, the family trait of bloody-mindedness coming to my rescue, until after almost two years of practising I made my first perfect globe. It's a really intense task. To this day, my favourite part of the production process is putting on the very last gore. Only then can you breathe a sigh and give it to the painters. Up to that point it's just some bits of map on a sphere.

Today, a new apprentice globemaker in the studio can expect to spend at least six months – and up to a year – learning how to master this one technique in our studio. I look for people who have the ability to learn a lot of the process themselves through observing the more experienced people in the team, are prepared to put in endless practice and, of course, like when I was starting out, can improve through trial and error. Working with your hands every day requires a passionate commitment to the task, and our work is so delicate and precise that it needs a certain personality. Globemaking takes patience, persistence, determination and stubbornness. After this initial learning period, an apprentice should be able to produce a small globe that will be good enough to sell; it takes a lot more practice before they can successfully tackle the larger globes, each one being a major milestone with the globemaker more excited than even the most excitable customer.

While modern developments in printing were helpful to me, on reflection they might well have been the reason that the art of

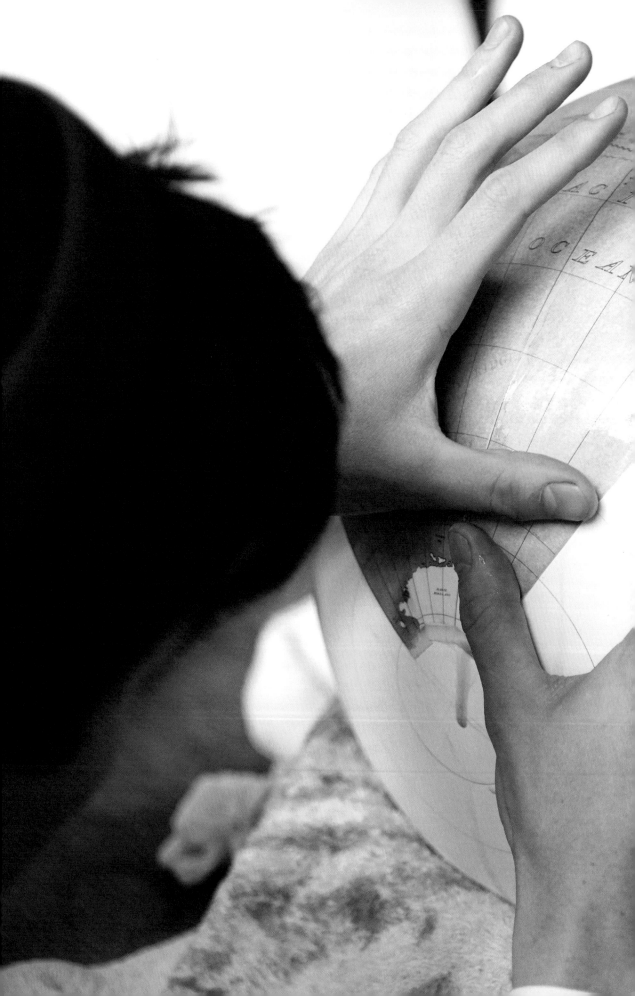

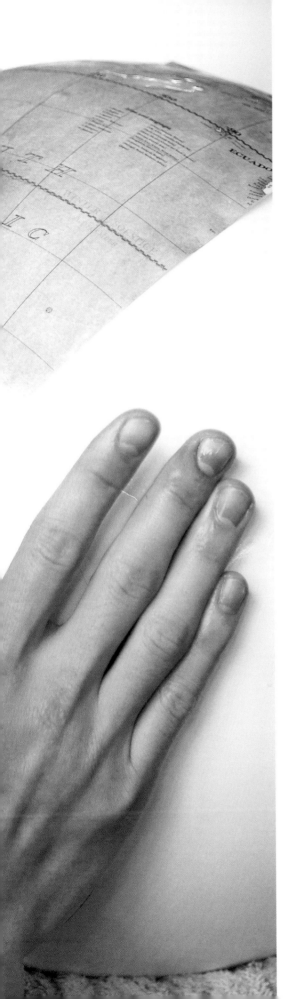

globemaking suffered such a downturn in the twentieth century. With the advent of full-colour printing, there was no need to employ a team of skilled painters to embellish and colour the maps, and with the arrival of competitively priced globes, it seems there was a race to the bottom. Previously, globes had been phenomenally expensive, more costly than many of the masterpieces that adorned the walls of the stately houses they sat in. Once machines able to steam-press or vacuum-shrink the map onto the sphere had taken over, globes became functional objects rather than the costly masterpieces of yore.

I had by now jettisoned hundreds of spheres into skips, along with the tip of my finger from a rogue scalpel blade, but I was finally happy with the end results of my labour and producing globes that I could sell. I fulfilled my first order – to the patient librarian in Brisbane – and on top of Yinka's exhibition globes and the challenge of making an even larger globe, the Churchill, on which more later, I now had a few more orders for my fifty-centimetre terrestrial model.

Sorry, but Dad would have to wait a bit longer for his birthday present. I had bills to pay.

6

THE HUNT FOR CHURCHILL'S GLOBE

While I was still wondering what to do about the dilemma of getting Yinka Shonibare's eighty-centimetre spheres through our narrow shop door in Kynaston Road, another prospective client, an American collector with whom I had been discussing an even larger 127-centimetre (50-inch) globe for several months, green-lit that project. The client had mentioned that his inspiration was a vast globe which had once been in the possession of President Franklin D. Roosevelt, and which was now on display at the Presidential Library and Museum in Hyde Park, just

'WINSTON CHURCHILL, HE TOLD ME, HAD AN IDENTICAL GLOBE. THIS WAS NEWS TO ME'

north of Poughkeepsie, New York State. Winston Churchill, he told me, had an identical globe. This was news to me. After some online research, intrigued and unable to mutter the word 'no', I agreed to the challenge. I also resolved to seek out and examine Churchill's globe for myself.

I wrote letters to Downing Street, Buckingham Palace, the Cabinet War Rooms, Blenheim Palace and Chartwell House, the Churchill family's former home in Kent, now a National Trust property, enquiring about its whereabouts. Within a week I had received replies from everyone – including a call from Downing Street the morning after I had sent the letter. Given the Churchill connection, there

THE CHURCHILL– ROOSEVELT GLOBES

For Christmas in 1942, the United States Army presented an enormous printed globe to President Roosevelt at the White House, where it took pride of place next to his desk in the Oval Office. At the same time, British Prime Minister Sir Winston Churchill received an exact duplicate of it in London, accompanied by a letter from General George C. Marshall, the US Army's chief of staff, ending, 'I hope that you will find a place at 10 Downing Street for this globe, so that you may accurately chart the progress of the global struggle of 1943 to free the world of terror and bondage.'

With each having a diameter of approximately 127.5 centimetres (just over 50 inches) and weighing over 340 kilos (750 pounds), they are possibly two of the largest printed globes ever made. The diameter was chosen because it is one ten-millionth of the size of the Earth. They were also the biggest globes that could be used without resorting to a ladder or placing them in a pit.

Work on the globes had begun a year earlier, shortly before the Japanese attack on the Pacific Fleet at Pearl Harbor on 7 December 1941. Four days later, Germany declared war on the United States. Colonel Donovan of the Office of the Strategic Service (the forerunner of the CIA) in Washington commissioned the globes with a specific purpose: the Allied leaders would have identical globes, the largest and most accurate ever made at the time. Apparently, the biggest printed globe then in existence was an English-made 30-inch globe that was badly out of date. For these new American globes, the nation's leading cartographers would compile the specially commissioned map, so it would be highly accurate and include many more up-to-date place names.

Fifty leading geographers, cartographers and draughtsmen worked for many months compiling, charting and checking the 17,000 place names included on the Churchill– Roosevelt globes – ten times as many as would ordinarily have been involved with globes of this size. If some of the place names in the map of the United States seem obscure, this is because everyone who worked on the map made a point of including their own home town.

With their matching printed globes, Churchill and others could reference them simultaneously to gauge distances by sea, a critical consideration in allocating limited shipping resources and formulating war strategies. In reality, however, the gift of the globes was a simple PR exercise, an important weapon in modern warfare. Despite the size of the globes, the map was simply far too small for any practical use.

was a lot of excitement about the globe and interest in my own globemaking endeavours, but ultimately no one knew where it was.

Eventually, months later, after seemingly exploring every avenue, I received an email from the Cartographic Society of Great Britain confirming that Churchill's globe was at Chartwell House – definitely. Yet, having spoken with Chartwell, I had been told they definitely didn't have it. So, I made another telephone call, and the same well-spoken man assured me there were no globes amongst the treasures at the property. I enlisted the help of author Simon Garfield, whom I had recently met, and who was following my progress on

The Chartwell staff subsequently allowed me to return on one of the museum's closed days to inspect the globe in detail. Churchill's globe had been at Chartwell it seems since the war. His grandson recounted to me later that it had worked perfectly well in the early 1960s, though when I examined it, it was showing signs of age. Due to years of touching, globes have a habit of wearing away at the point on the map where their owners reside, hence much of the UK was missing. I believe it has at some stage had conservation work done, but there is a bit of debate in museum circles about the relative merits of restoration and conservation. I agree that you can't just make Churchill's globe as

'WE'RE VERY WORRIED THAT OTHER MUSEUMS HAVE THEIR EYES ON THE GLOBE, SO WHEN PEOPLE CALL, WE TELL THEM IT'S NOT HERE'

the making of the behemoth for the American client for his new book, *On the Map*. We drove together to Chartwell, and as we joined the queue to enter the house, I recognised the voice of one of the volunteers working the entry gate. Again, I asked the well-spoken man, 'Are there any globes here?'

'Churchill's globe!' he exclaimed. 'Of course, young man!' he enthused (I am in my forties), adding, 'It's in his studio, over there. It's absolutely magnificent.'

'Are you sure?' I asked. I explained that I had called earlier in the week and was assured that there was no globe in the collection and that the voice I had heard on the phone sounded like his.

'Aha!' he said. 'We're very worried that other museums have their eyes on the globe, so when people call, we tell them it's not here. Don't let on ... Sorry!!'

new again – far easier just to make a replica – but it is gradually falling apart, degrading, and at some point it will either be beyond any sort of repair or will require extensive and expensive restoration. Eventually, someone must make a decision. Nevertheless, it's inspiring to view, and the history gives it such importance.

The prospect of replicating a globe of such vast size was daunting, the design challenges immense. First, which material would be most suitable for a globe of this scale? GRP (fibreglass) seemed the obvious choice, as it is strong and lightweight. Second, how could I make an accurate mould to cast the sphere? When I mentioned my requirements to a GRP fabricator, my eyes lit up when he mentioned that his pattern maker also made body components for Formula 1 racing cars. The precision of these cars was exactly what I wanted, and so I commissioned a pattern and mould. For the

pattern, a CNC (computerised numerical control) machine revolved laboriously around a polyurethane block for two days, turning out a perfect 127-centimetre hemisphere. Sad to say, the video that they sent me of this process did not go viral. From this pattern, we then created a mould from which to create our hemispheres. Locating points were built into the mould's internal rim, allowing two hemispheres to be perfectly aligned and bonded together.

Once the join line had been made good, I then had to work out how to attach the huge gores to the sphere. I was working with gores cut at the equator, but even so each of the sections of printed paper was 1.2 metres long. My first problem was that when I drenched one of these in water, it became so heavy that I had to lift it super-carefully, otherwise it ripped under its own weight. Then once I laid the paper on the globe, I had to work at breakneck speed because there was a limited amount of time before it stuck fast to the glue. I had to keep the gore moving until I had positioned it correctly, but then ensure that it didn't move any further before I started on the next section, so I enlisted Meredith, my new painter, to hold it in place or tell me it had slipped so I could reposition it. In the end, it took several weeks to complete this part of the process. Today our apprentice globemakers learn to apply one single gore north to south; successfully laying their first gigantic gore on a Churchill with the help of a colleague is a seminal moment in their careers.

Once the gores had dried on the globe, Meredith and I spent a further six to eight weeks painting the map. This was another gargantuan task. The sheer scale and weight of the globe meant that every time we shifted it to begin work on another section, we had to enlist the help of several other people, but once we added the last brushstroke and stood back to see the completed globe, it was undoubtedly a beautiful work of art.

time, the PM exclaimed, 'Where the hell have you been, Davenport!?' The globe, however, had arrived just in time for Christmas, and pictures of the two leaders posing with their matching globes circulated as part of wartime propaganda.

I also talked to Bradfute's elderly mother, who was delightful. She spoke at length about the whole project and confided that on the return leg of the journey, the captain of her

'AT THE FRANKLIN D. ROOSEVELT PRESIDENTIAL LIBRARY AND MUSEUM IT WAS DISCOVERED IN STORAGE'

Much later, after we had made and shipped several more 127-centimetre globes, I received a call from an American called Bradfute Davenport. He had read about Bellerby & Co. and our Churchill globe in Simon Garfield's *On the Map*. Bradfute's father, he explained, was Captain B. Warwick Davenport, who in December 1942 had been responsible for the safe delivery of the globe from General Marshall to Prime Minister Churchill.

Bradfute recounted the whole story, how the globe had started its epic sea voyage from Newfoundland, but that the ship had to turn back due to bad weather. It was then diverted to Brazil (some diversion!), so when his father finally presented it to Winston Churchill, who had known of its impending arrival for some

husband's ship was asked if he could help someone out with a lift. It turned out that this was General Charles de Gaulle, who needed to get to Africa, perhaps in anticipation of the Allied Casablanca Conference of January 1943, led by President Roosevelt and Prime Minister Churchill. This happened to be on the ship's route, so the captain happily invited the general on board. Mrs Davenport said that even in the 1980s, when she and her husband had tried to find Roosevelt's globe, it had been largely forgotten. At the Franklin D. Roosevelt Presidential Library and Museum it was discovered in storage, and no one there knew anything about its history. On realising its significance, the museum hastily made some rudimentary repairs to the globe and put it on display.

7

THE MERIDIAN AND ENGRAVING

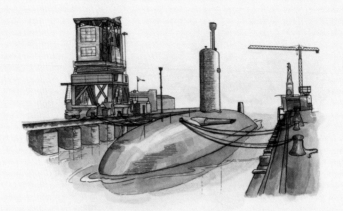

WHERE ARE ALL THE FOUNDRIES?

My father was a naval architect at BP Shipping, designing and overseeing the construction of tankers and emergency support vessels for the oil industry. He worked all over the world, mainly in Europe's shipyards – in Scandinavia, the Low Countries and Split in Croatia (part of the former Yugoslavia) and in and around the UK, especially on the Clyde estuary in Scotland. We often visited him, spending entire summers in Belgium and Sweden. We stayed in one hotel so long, my siblings and I became the bartender's little helpers.

I remember a visit to Scotland in 1976 (despite there being a record-breaking drought in England, it rained every day) and being so excited as I was shown around a submarine in dry dock at the Scott Lithgow shipyard in Greenock, just outside Glasgow. I was eleven years old so quite small at the time, but the space, or lack of it, on that submarine and the amazingly efficient utilisation of every inch of it, is something that has lived with me to this day. From an early age I loved engineering, and I fired all sorts of questions at the submarine's captain, which presumably for security reasons largely went unanswered. I remember being most impressed by the fact that submarines have two skins, external and internal, to withstand the enormous pressure deep below the surface of the sea.

I was amazed by the vast scale of the heavy industry in the shipyard: numerous fiery foundries, towering cranes and soaring scaffolding,

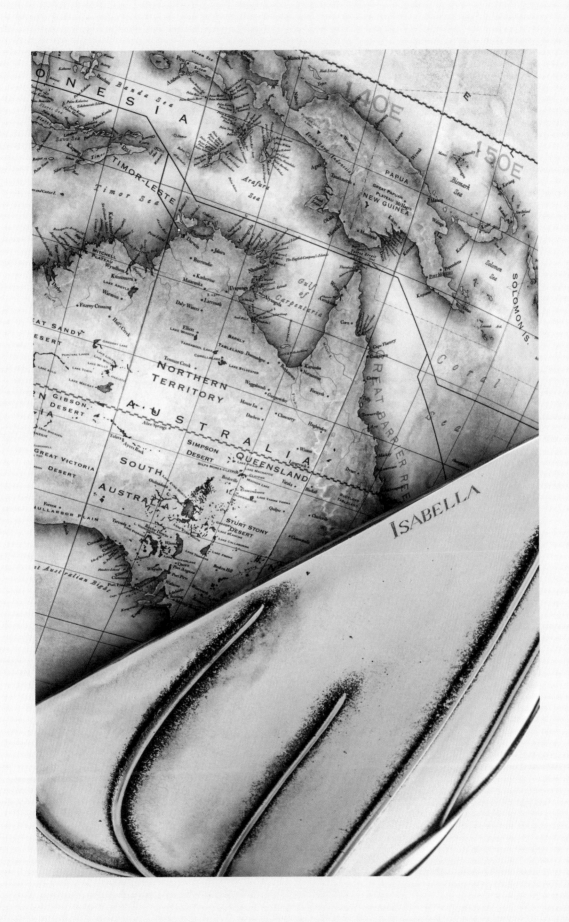

with welders' sparks flying off the skeletal ships in production, oversized machines and unique shipyard vehicles lumbering purposefully around the site.

This was back in the 1970s, when the UK still had a considerable manufacturing sector, albeit much of it loss-making and in terminal decline. Shipbuilding had once been a big employer but, like many industries in post-war Britain, by the early 1980s UK firms could no longer compete with the low wages, skilled workforces and high productivity of overseas companies, and the South-East Asia, especially South Korea, now dominates shipbuilding. Sadly, only a few shipyards remain operational in the UK, mainly repair and servicing yards, and military shipbuilders, where national security demands that certain vessels be manufactured in the UK.

The extent to which UK manufacturing has declined was not really something that had previously affected me. I like to support the local or national economy, but as a consumer you can only do so much. Now I discovered the consequences of our dwindling industrial sector. A traditional terrestrial globe consists of a globe surrounded by an engraved meridian ring, usually cast in brass and framed by a painted horizon band set within a wooden table. The meridian is fitted vertically within the wooden base or table, and can be moved through any angle so the globe, which is held within it by an axle, can be spun as required. And while today they mainly perform a decorative function, they once served a navigational purpose. I needed to find a foundry to make the meridian rings for our globes.

THE PRIME MERIDIAN

Longitude lines run between the North and South poles. The Prime Meridian is defined as 0°00'00" longitude, and everywhere on Earth can be measured east or west of this line, with half of our planet east of the Prime Meridian and half west. While the equator does the same in a north–south plane, its position is fixed as it is determined by the Earth's axis. The Prime Meridian is, however, arbitrary. When Mercator made his globe in 1541, he had it run through the Canary Islands, and on later globes it goes through the Azores. Later, longitudes were marked from 0° to 360°, and this system was used by navigators well into the eighteenth century.

Not until the 1884 International Meridian Conference in Washington, DC did the Prime Meridian become fixed at one point – through the Greenwich Meridian, south-east London – and the longitudes become universally marked as being 180°W to 180°E. Even then there were some dissenters to the nuanced decision that it should not pass through a major continent, so as not to be advantageous to one. The position needed to be an established observatory on land to ensure accurate measurement, so there was little alternative to London.

Today all the meridians on our globes are hand-cast in solid brass and hand-engraved. At the outset, however, I neither knew precisely what I wanted, nor who I should approach, nor anything about metal casting, but there seemed no reason to believe that this essential piece of globe construction would prove problematic. But I didn't exactly know what was achievable in casting, nor even that CNC precision metal machining existed or might be an option. Using computer programming instructions, this method of machining allows a metal object to be cut and shaped to precise specifications without manual intervention by a machine operator. In my naivety, I almost imagined that I would find a couple of foundries with a ready stash of meridian patterns.

Alas, my initial online search was not encouraging, but I figured that in 2008 many foundries were probably not particularly internet-savvy and had simply not set up websites, so I printed a list of companies from the UK Cast Metal Federation website and called at least 150 foundries. I pretty much exhausted their entire directory, but half of the numbers were no longer in service. To most of those who answered, I had to explain what a meridian was. I drove up to the Black Country to visit a few foundries. Many of those I spoke to suggested using the lost wax technique, but none seemed confident of casting to the scale I wanted, or could suggest any solution. Months went by with little headway.

In the end I commissioned a couple of small foundries to cast some prototype meridians, but when they arrived, they were not perfectly round, had cracks due to overfast cooling (a rookie mistake which didn't inspire confidence), and I had to pay for all of them. In advance. No refunds for shoddy work.

THE LOST WAX PROCESS

Lost wax casting, also known as cire perdue, is a method used since around 4500 BCE for casting metal alloys or moulding glass. A sacrificial wax replica of the item to be cast is made (every time, so if we need 40 finals, 40 sacrifical ones are made), then plaster is poured around the wax model inside a cask. Once the plaster has set solid, the cask is heated at high temperature in a kiln so the wax melts away, leaving only the plaster mould. The molten metal or glass is poured into this mould, filling the space left by the wax. The method allows you to capture extremely fine details in both metal and glass.

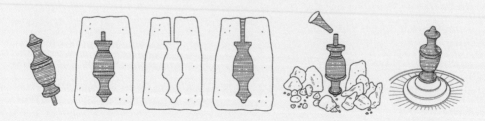

My initial idea was to cast the latitude scale into the meridian. I was struggling to find anyone to engrave meridians by hand and had seen one or two meridians with sand-formed numerals, which I thought might work if I could improve the quality. Doing this, I could avoid the cost of sourcing and commissioning a hand-engraving. There were nevertheless numerous other construction variables and decisions that I needed to make, and they were all going to cost a small fortune if I got it wrong. So I looked more closely at how I should design the meridian, and how it would fit into the wooden base.

When set in the table mount, the meridian is held in place by two notches on either side, with a cradle beneath to keep it vertical, house the bearings and support the combined weight of both the meridian itself and the globe. The latitude scale on the meridian should line up perfectly with the latitude map lines on the globe, which are all derived from its central axis. However, if you set the globe inside the centre of the meridian, the thickness of the meridian, usually 10–15 millimetres, means that the longitudinal markings do not line up with either side, so instead the axle of the globe is mounted half within the meridian and half outside and secured by a clamp,

which protrudes out from the meridian.

The axis of the globe is thus centred to run flush with one edge of the meridian, lining up perfectly with the latitude engravings on that side (the engravings on the other side are just for show). The notches in the table to hold the globe and meridian in place are therefore also slightly off centre to allow for the fact that the meridian sits further out on one side of the globe than the other. Wrongly placing the globe and meridan within its supporting table will lead to the globe making contact with its table as it spins, likely causing damage.

Evidently my planning had to be spot on. I need not only a meridian, but also a whole host of brass fixings: a clamp and small screws to secure the axle to the meridian and saucer-like cup washers to fit over the axle at each end, with bespoke nylon bearings inserted to mitigate any squeaks when you spin the globe.

When I was fixing up my Aston Martin DB6, I had spent hours with my mechanic friend Paul, who repaired Land Rovers from his workshop under a railway arch in London's East End. I learned a lot from him, not least that engineering was way more flexible than I had imagined. Having successfully fixed the suspension, drive train and axles on the 130-mph car (and afterwards driven in it on

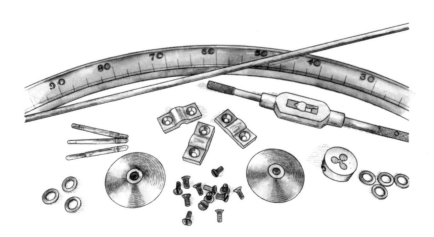

'I WAS STRUGGLING TO FIND ANYONE
TO ENGRAVE MERIDIANS BY HAND AND
HAD SEEN ONE OR TWO MERIDIANS WITH
SAND-FORMED NUMERALS, WHICH I THOUGHT
MIGHT WORK IF I COULD IMPROVE
THE QUALITY'

Molten aluminium being poured into moulds composed
of super fine sand and gasses

a German autobahn),
I was confident that my
design for fixing the meridian
and globe together would work.

After much searching, I had
found a highly skilled man called John,
who had ran Manor Foundry in Derbyshire
casting anything from engine parts to
decorative bronze swords. It was clear from
the outset that John was intrigued by the project;
he took me under his wing and helped me to work
out the best meridian and fixings designs and how
to fit them in place around the globe. Finally, I had
found someone who didn't just talk about how much
things would cost the whole time. Nothing to pay until
completion.

Foundry work is intoxicating. It really is like going back
to medieval times, with neat piles of ingots of different
metals, mounds of sand of different colours and textures,
furnaces roaring, and everyone looking like they have spent
the last year down a coal mine without washing. Smoke
billowed around the eaves in John's industrial unit, and
patterns of all his previous projects adorned the walls.
While probably not ideal for the lungs, the fumes were
fantastic for the soul.

I spent several long days there watching the whole process
from start to finish. From his extensive experience of the
casting process, John believed the best idea would be to
cast a single 20-centimetre-high hollow oil-drum-
shaped piece, from which we could make multiple
meridians at once. He created the mould, then
poured the molten metal into it until it was full,
using huge ladles that glowed red. There are
many variables to consider, shrinkage being
an important factor. Brass is an alloy of
the metal elements copper and zinc,
which have different molten
temperatures so needed to be
added to the furnace at
different stages and

in specific quantities
along with gases to ensure
the mixture flowed through the
mould. Once he had extracted the
cast and left it to cool, John clamped the
hollow brass drum to a lathe, smoothed off
the edges and used a parting-chisel machine
to create ten perfect meridians.

Had I known it were possible, I could have
simply machined sheets of brass from the outset,
but there is something romantic about having the
meridians cast in a foundry, and besides this way we
minimised waste. John then arranged for the small fixing
pieces, cup washers and clamps to be machined or lost
wax cast. Initially, we also made several attempts to sand-
cast the meridian numerals, but they lacked clarity, so for
the first few, until I found a hand-engraver, John had them
engraved by CNC machine for me.

Now almost finished, the meridians still had machining
marks in the metal from the parting chisel, tiny and not
noticeable to the touch, but definitely to the eye. I had to
polish these out by hand. Another technique to learn.
Thankfully, it is marginally better than sawing plaster
hemispheres. I bought a 1950s polishing wheel, which is
essentially a motor in a cast-iron housing, driving a disc-
shaped mop – a bit like a giant angle grinder, but instead
of a titanium blade it has a soft padded wheel about the
same size as a roll of gaffer tape, positioned at waist
height. The polishing mop is used in conjunction
with polishing soap, which comes in varying grades
of abrasiveness, and the mops themselves also vary
in their roughness.

I held the meridian firmly against the mop,
and the offending machine marks gradually
disappeared. In the process, soap and
mop pieces flew off in all directions
at speed, the soap turned dirty,
the metal started to heat up
from the friction, and
my gloves turned

grey, along with any uncovered skin or hair. But eventually, after an hour or so, a beautiful polished meridian resulted. Polishing the little metal fixing pieces was even more fun; they were too small to grip with gloves, and the metal got so hot that it was difficult to keep hold of them. Sometimes I just had to let go, and the fixings flew off the spinning mop, thwacking against the walls and floor. I found it useful to wear protective eyewear when performing this task.

Tap and Die Company, specialists in this niche field, so I could source replacement cutting bits quickly.

Finding a hand-engraver, however, took several years. Indeed, it seems this may well be an existential point in time for this important craft. Many of the jewellery shops in Hatton Garden and in the West End used to have engravers on their staff; nowadays, though, almost all engraving is done by machine. Today

'FOUNDRY WORK IS INTOXICATING. IT REALLY IS LIKE GOING BACK TO MEDIEVAL TIMES'

Next, I had to learn how to tap and die. This allows you to create threads within a solid piece of metal into which you can insert screws. Another first for me. After drilling a pilot hole, it's a careful process of turning a screw-shaped cutting tool slightly larger than the drilled hole to create the thread. It sounds simple and is, but the cutting tool is so thin, and despite being soft, brass is still a metal, after all, so the bit was prone to snapping and would end up lodged in the meridian. Removing the broken bit from the softer brass without destroying the expensive meridian was a four-hour process. Thankfully, I found a small shop just three miles away on a Tottenham back road called the

we collaborate with a wonderful engraver with a small studio in Clerkenwell because, for me, a handmade globe absolutely needs a hand-engraved meridian.

I find newly polished brass finishes a little bright, so the final piece in the puzzle was to gently age the brass to give a more distinguished, timeless look. I found some antiquing fluid in a hobby shop in Warren Street in central London, but it had little effect. Alternative ageing methods include burying the brass and waiting, but time was a little pressing. Happily, the man in the workshop of the Warren Street store agreed to make up a much stronger solution especially for my purposes, and the problem was solved.

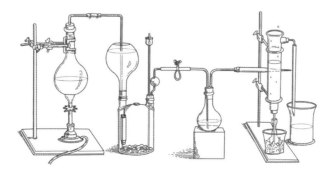

8

THE COLOURS OF THE EARTH

PAINTING THE GLOBE

While refining and perfecting the art of applying the paper gores, my thoughts had turned to the art and aesthetics of painting the globes. Remembering Blaeu's seventeenth-century celestial and terrestrial globes, I resumed my historical research.

The simplest way to make a globe is to construct a sphere and paint it. The earliest globes would have been made of wood or metal, with the celestial or terrestrial map painted directly on by hand. Later on, hollow globes were made of thin sheets of metal, giving greater size and flexibility.

Initially, makers pasted blank gores onto the sphere to create a more forgiving canvas for the hand-painted map and lettering. These are called manuscript globes and due to the low production levels of globemakers, manuscript globes continued long after the invention of the printing press, at which point whole maps could be printed as gores. A silversmith or skilled engraver would etch a map on copper plates before printing using a process known as intaglio, from the Italian word for 'carving'. In intaglio printing the etched plate is coated with ink, then wiped to leave ink only in the incised depressions, before being run through an etching press, in which dampened paper picks up the ink to create the printed image. Copper is a soft metal, so the plates lose their clarity relatively quickly; smaller print runs were therefore common. The effect, though, is very satisfying, with an intense character to the image. The globemaker then pasted the printed gores onto the globe and finally the painter would add colour.

It was at this point that the globemaking craft became assimilated with the printing and publishing industry. Globes were after all now printed just like books, and since this time each edition has been referred to as a 'publication'. And as in book publishing, copying the map from a rival's globe is plagiarism.

The golden age of the printed and then hand-painted globe coincided with the age of European expansion, reaching its peak at the beginning of the seventeenth century. In this period, as astronomical, geographical and cartographical knowledge developed apace, globemakers too were inspired to experiment and refine their art. In turn, the proliferation of printing presses made it possible over time to produce more globes at a less than exorbitant cost so they became more affordable to a greater number of people.

Nevertheless, the acquisition or commission of a globe was still the preserve of the aristocracy and the affluent merchant class. Because of the delicate and time-consuming nature of the work, a budding globemaker probably would have required considerable financial backing. Globes were therefore prized symbols of status and prestige.

Studying these venerable antique globes, it was striking to see how little the methods of manufacture had changed from the mid-sixteenth century until the twentieth century, albeit there is always a mystery about the exact construction and methods because so much is hidden under the surface – it was only in the last century that the rot set in. I knew that I had high aspirations but I did not want to simply reproduce some sort of cheap faux-antique facsimile. Instead, my ambition was to produce a handmade globe that felt classic yet at the same time unusual, relevant and contemporary.

I come from a line of keen artists. My grandmother and my mother both loved painting with watercolours; my grandmother even taught it for many decades until well into her eighties. I have several of their paintings, although they are stored in my attic because, sadly, I just don't share their enthusiasm for this medium; I always say that I don't like the imprecision of the application, although more likely I don't care for watercolours because I have never been very good at painting with them. However, in collaboration with the crispness of the cartography on a globe, watercolours acquire another dimension, allowing you to build up a rich colour patina over many layers without obscuring the text. It really is a perfect match.

Watercolours were no doubt used on the finest old globes for this reason; indeed, I would go so far as to say they could have been invented for globemaking had they not been conceived centuries earlier than the first painted globe. Globemakers must surely always have planned to paint their globes with watercolours; they knew their creation would have pride of place in the purchaser's house, so beauty was paramount. We might admire the look of these old globes now, but when they were made, they were positively revered.

KYRGYZSTAN
Bishkek
Almaty
Tashkent
Chirchik
Osh
Fergana
Andijan
Khudzhand
Gagarin
Dzhizak
TAJIKISTAN
Samarkand
Dushanbe
Qurgonteppa
Kulob
Khorog
Feyzabad
Konduz
Kabul
Gardeyz
Qalat
HANISTAN
Lashkar Gah
Qandahar
Zhob
Quetta
Dalbandin
PAKISTAN
Khuzdar
Shikarpur
Sukkur
Indus
Panjgur
Hyderabad
Gwadar
Ormara
Karachi
OMAN
Chitral
Asadabad
Peshawar
Islamabad
Rawalpindi
Lahore
Faisalabad
Multan
Bahawalpur
Kashgar
Tokay
Yecheng
Taxkorgan
Hotan
Keriya
Yeyik
Qilan
Bachu
Mazartag
Akto
XINJIANG UYGUR ZIZHIQU
Mani
Lungdo
XIZANG ZIZHIQU
Gerze
(TIBET)
Urmai
Xiongmei
Kunglung
Xigaze
Zhongba
Yalaixiang
Everest
SIKKIM
K2 8611m
LADAKH
Nanga Parbat 8126m
Srinagar
Leh
JAMMU AND KASHMIR
Jammu
HIMACHAL PRADESH
Amritsar
Chandigarh
UTTARAKHAND
PUNJAB
HARYANA
DELHI
New Delhi
UTTAR PRADESH
Dhangadhi
Pokhara
NEPAL
Kathmandu
Birgunj
Janakpur
Lucknow
Kanpur
Allahabad
Varanasi
Patna
BIHAR
Gaya
JHARKHAND
Ranchi
Agra
RAJASTHAN
Jaipur
Pushkar
Jhansi
Mirzapur
Jodhpur
Kota
Udaipur
Bhopal
Sagar
Jabalpur
Raipur
CHHATTISGARH
Sambalpur
Cuttack
ODISHA
Kolkata
Ahmadabad
Indore
MADHYA PRADESH
GUJARAT
Morbi
Rajkot
Bhuj
Jakhau
Okha
Porbandar
Vereval
Bhavnagar
GULF OF CAMBAY
Surat
Daman
INDIA
Nagpur
Kanker
Jagdalpur
MAHARASHTRA
Chandrapur
Nizamabad
TELANGANA
Hyderabad
Rajahmundry
Kakinada
Vishakhap
Machilipatnam
Guntur
Krishna
Pune
Mumbai
Satara
Solapur
Bankot
Miraj
Bijapur
Belgaum
Ratnagiri
KARNATAKA
Bellary
ANDHRA PRADESH
Nellore
Chennai
Panaji
GOA
Karwar
Davangere
Kolar
Bengaluru
Vellore
Puducherry
Coondapoor
TAMIL NADU
New Mangalore
Mysore
Coimbatore
Tiruchirappalli
Kozhikode
Madurai
Jaffna
Ernakulam
ARABIAN SEA
Island

Meanwhile Chiara Perano, a friend of Jade's obsessed with astrology and mythology, had been designing a celestial globe, mapping the stars and drawing all eighty-eight constellations by hand. She also decided that my original basic cartouche was not suitable for her celestial globe, and she quickly came up with a much better design.

In the early years of Bellerby & Co., my approach to publicity and marketing was a little scattergun. Finding the correct person to contact at publications for editorial content was far from straightforward. I just fired off the odd email here and there, and occasionally the employee handling the info@ or press@ account would pass it on to the editorial team. Sometimes this miraculously resulted in some publicity for Bellerby & Co. globes, such as a tiny feature in *House and Garden* magazine.

Just as Chiara was finishing the first Bellerby & Co. celestial globe, the Perano Celestial model, David Balfour, the property expert on Martin Scorsese's Oscar-winning movie *Hugo*, saw the House and Garden piece and commissioned me to make four globes for a scene in the film, one of which was to be a celestial globe in two pieces; they were going to film the scene in a clockmaker's studio, so our globes fitted the bill.

The deadline for the *Hugo* globes was ridiculously tight – filming was due to start in June 2010, and I had to build in extra time for their in-house approval. And I was still learning many of the processes and practising only on 50-centimetre globes; the commission was for a 40-centimetre celestial globe and three much smaller terrestrials. I worked into the night for weeks for next to nothing – I was just excited to be asked.

After finishing the four globes for the movie set, I realised that my painting talents – and patience – were never going to be up to the job. I had taken several days to paint the company's first ever globe and then someone (nameless) had dropped it, making it unusable. This didn't do that much damage, but you can't easily repair a globe from the outside, and besides a repaired globe isn't a new globe. This was a clear sign. I painted a few more, then took on my first proper employee, Meredith 'Dith' Owen, as resident painter.

WATERCOLOURS

Prior to the late eighteenth century, artists would have made their own watercolour paints, keeping their recipes a closely guarded secret. It was not until 1780 that William Reeves invented the square soluble pans that we take for granted these days. At this stage of my project, I had plenty of time on my hands so I decided to make my own watercolours just as the early masters had done. I duly went to L. Cornelissen & Son, a wonderful artists' supply shop in Bloomsbury, central London – which wouldn't seem out of place in Diagon Alley – and bought various pigments, ox gall, honey, gum Arabic, along with a glass pestle with which to grind them. Feeling like an eighteenth-century artisan, I could not wait to get home and get mixing.

been doing it for generations and has a lab full of chemists to prepare the formulations, it's a false economy to try to compete.

I applied the background colour by gently but quickly dipping the gores in large troughs of dissolved watercolours. Combining several colours in each trough created the desired palette, and I dipped each gore in several mixes to achieve a subtle earthy, ochre tint.

The next issue I struggled with was the space needed to allow the sets of freshly painted gores to dry. Rather than patting the paper dry and potentially flattening the colour, I wanted to leave the washes on the gores and allow them to air-dry naturally, which produced a richer texture.

'FEELING LIKE AN EIGHTEENTH-CENTURY ARTISAN, I COULD NOT WAIT TO GET HOME AND GET MIXING'

What a palaver. I tried many times but could never get the paint to bind properly. All that kit and those unused pigments are still sitting somewhere in the studio, stuffed in an old wine cooler. Instead, I went to our local art store and bought several tubes of pre-prepared watercolour paints; more expensive but a better use of my time. I'm not sure if it would have worked out cheaper to mix my own paints, but sometimes it's just better to accept that when someone is producing something to a very high standard, has

A 50-centimetre globe requires almost two square metres of floor space when the gores are laid out flat, and each gore would remain wet for several hours – I needed somewhere to put them all and I just didn't have the space. So I ran some string between the walls at the shop and hung up the quickly multiplying gores with clothes pegs; then I and laid more across the heated kitchen floor at home where the cats would run all over them. I'm pretty sure Dad's globe has a few paw prints on – they knew he didn't like cats so they wanted to leave their mark!

Once I had attached the tinted gores to the sphere, I could tackle the final painting stage. This is the most detailed and one of the most time-consuming painting jobs, but also the most satisfying as it brings the globe to life. I used a very fine paintbrush and began to paint around the seas and lakes to add contrast between the land and water. As no doubt many artists know, I quickly found that watercolours tend to leave streaks at the edge of each stroke and subsequent washes don't properly amalgamate. However, I discovered that certain pigments (or at least some of the ones I was using) have an unexpected property: when mixed with other pigments, they have the effect of chasing them away, seemingly due to differences in the surface tensions they create in the water. If I was careful with the ratio of water to pigment, this allowed me to achieve a perfectly graduated fade away from the coastlines. After seeing Blaeu's exquisite masterpiece at the Maritime Museum, this was the effect I wanted and how I imagined my globe would look.

Of course, when painting a globe there is another obvious issue which is impossible to ignore but unique to globemaking. I find it enough of a struggle working with watercolours on a flat surface, but painting onto a sphere is an altogether different challenge. How do you apply a pigment which is essentially floating in water onto a sphere without it running everywhere and leaving particles of pigment along its journeys? In order to avoid this, I had to move the globe constantly to ensure that the section I was working on was uppermost. To get the desired fading effect, my method already involved painting in tiny sections, pre-wetting each precise area with clean water, so that I could add the pigment with less liquid. In this way, the pigment dispersed only as far as the paper had been wetted. The many small islands around the seas off Canada, Scandinavia and Malaysia took hours to paint, but eventually the transformation was complete, bringing definition to the Earth's coastlines.

MATCHING PAIRS OF GLOBES

In Christian Europe, it wasn't until the early sixteenth century that a Nuremberg burgher, the former cleric turned printer, Johan Schöner, had the idea of making matching printed (from woodcuts) terrestrial and celestial globes. Early purchasers loved to have the set. To have this pair of scientific instruments in your home was a big status symbol as they were evidence of the owner's knowledge of both the world and the heavens. This was the Age of Exploration, with nations vying to conquer new lands and stake their claim on fruitful trading opportunities, but this was also a time before the discovery of any reliable methods of navigating at sea, and the globes served a useful purpose. Sailors depended on the positions of the stars to guide them across the treacherous oceans. Thus Schöner's matching pair of globes, first commercially printed and published in 1515, were not only attractive status symbols for a wealthy elite, but more crucially a celestial globe became an important, practical navigational tool.

All that remained was to protect the globe and its painted map. I felt it was important to apply any finish by hand, but again the sphere caused complications. The paper would quickly absorb the varnish, which would dry almost instantaneously, leaving streak marks – even worse, the brush would get stuck to the paper. It required speed, precision and applying just the correct amount of varnish. Too much and it would run, too little and it would dry too quickly and leave a rough finish. This was the finishing touch, and so much was at stake. Several years later, we discovered that varnishing alternate gores made the process much easier, but until that breakthrough this was yet another stressful moment. In addition, instead of varnish we now use modern protective coatings, especially for the globes that sit on roller bearings, which need a more durable finish.

Early varnishes, used to coat materials such as ceramics, wood and metal to make them more durable, waterproof and visually attractive, were developed by mixing a transparent resin (pine sap being a common example) with a solvent (often distilled pine sap) and a drying oil, and applying them by brush to attain a golden effect and hard finish. The ancient Egyptians were very competent in the art of varnishing. It also became common some time later in east Asia.

In China, the practice of lacquer work, a style of varnish application using resin from trees indigenous to east Asia like the lacquer tree and the wax tree, appears to have originated during the Neolithic period, and it has been suggested that artisans in Japan were well acquainted with the art of lacquering as early as 7000 BCE, during the Jomon period.

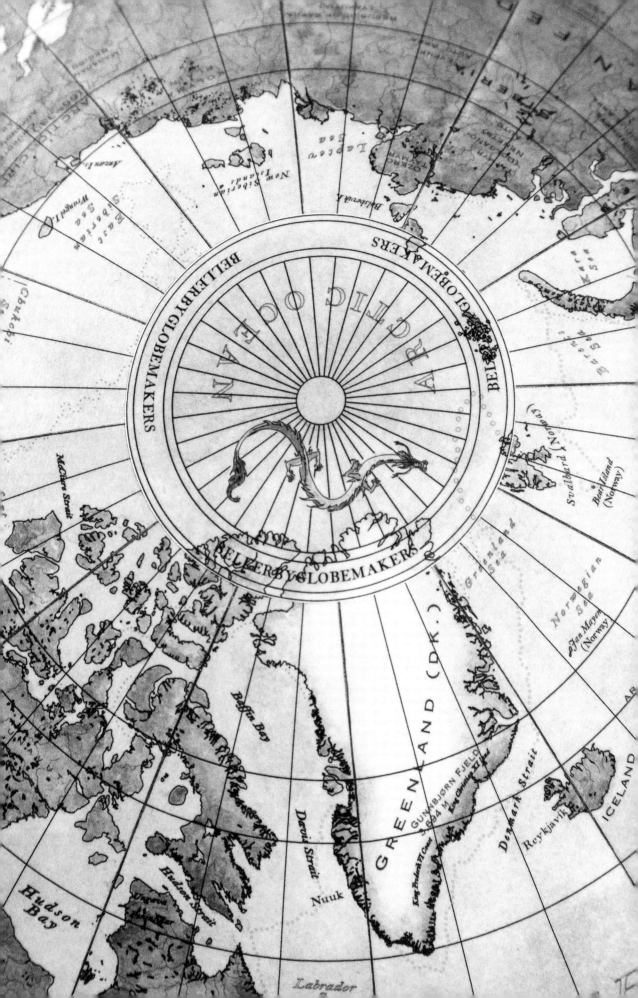

CELESTIAL GLOBES

While terrestrial globes are spherical maps of the surface of the Earth, celestial globes use the Earth as an imaginary centre of the universe to map the stars in the night sky. Celestial globes do not depict the sun, moon or planets, as their positions continually vary relative to those of the stars.

The ancient Greeks knew the world was spherical and made the first globes to depict their understanding of it. Their astronomers were quick to adapt the principle of the globe to mapping the skies, etching the constellations onto metal spheres. They thought that the stars sat on the surface of a giant sphere around the Earth, and that their constant movement every night and throughout the year was caused by this giant sphere slowly turning overhead.

Later, around 150 CE, the Romans depicted a 63.5-centimetre (25-inch) celestial globe in the Farnese Atlas, a marble sculpture of Atlas bearing the mighty weight of the heavens on his shoulders. It depicts the outline of forty-one constellations against a grid of circles including the celestial equator, tropics and ecliptic. This model, now housed in the Archaeological Museum in Naples, is thought to be a copy of a much earlier Greek globe of the constellations, based on Hipparchus' star catalogue of around 125 BCE.

The Greeks' astronomical theories quickly spread among Arabic, Persian and Indian scholars, who built upon many of the achievements of classical Greek science, refining concepts and the design of astronomical instruments, such as the celestial globe and the astrolabe. Islamic celestial globes usually featured the classical constellations, such as the Great Bear, Pegasus, Orion and the twelve signs of the zodiac, engraved onto hollow metal orbs.

A celestial globe can be constructed and viewed in one of two ways: as if the viewer is positioned at the imaginary centre of the Earth looking out, or as if the viewer is at infinity looking in.

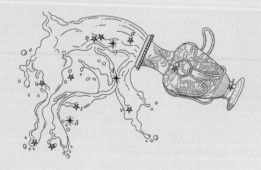

THE NEW SPACE: BOUVERIE MEWS, N16

By early 2011, space, specifically the lack of it, in the Kynaston Road shop had become an insurmountable problem. When I started shipping globes, I discovered that the front door was not big enough for a crate to pass though, so I had to check the weather forecast first, then construct a crate outside the shop. I would then arrange a pick-up with a courier company before I packed the globe inside and hoped that they would arrive when promised. Otherwise, I would have to reverse the whole procedure at 7 p.m. that evening and put the globe safely back inside the shop before I could go home.

And so in March, with the deadline for Yinka Shonibare's five 80-centimetre globes fast approaching and the order for a 127-centimetre globe from the American client bearing down on me, we moved to our new premises in Bouverie Mews, Stoke Newington. I had seen the building the previous year when a wishful-thinking estate agent had shown me the space, but at that stage I had not even mastered the process of creating a basic plaster-cast sphere, so it would have been foolhardy to take on the lease. It was cavernous. A vast twenty-square-metre ground-floor workspace, a dodgy little mezzanine at one end, broken windows, barely any power facilities, no heating and a bizarre pair of landladies who might have been teleported from the seventeenth century.

Still, given the scale of the outstanding commissions, drastic steps had to be taken. I had a huge door so I could easily get the large globes in and out of the building, but in truth it was far from perfect; the building dated from the 1950s and the owners had barely

spent a penny on its upkeep in the interim. They had however slashed the rent since I'd last viewed it, and it seemed about fair, but it was more than I could comfortably afford. It would be a struggle to get the deposit together, and I discussed options with Meredith, who was now painting for me three days a week. She introduced me to her brother, an artist, and to a group of art directors, fellow graduates from Falmouth College of Art. They were all looking for studio space, so I told the owners that while I loved the building, it was too big for me; I had plans to grow the company but not that quickly. Happily, they agreed to allow me to sublet, and the We Work-style studio was born.

We split up the space, erecting a Victorian spiral staircase to the mezzanine and sharing the downstairs area, though the building was still much bigger than we required and needed so much work. I set about making it comfortable and fit for purpose. It was a rent-and-repair lease, so as the tenant I was responsible for repairs. My landladies were indignant about any suggestion that they should be responsible for fixing the broken windows and leaking roof prior to us moving in; they expected me to sort all that out – and pay for it. Once I had taken possession, I discovered that the space had been used for illegal raves over the previous two fallow years; the last tenants had seemingly kept a set of keys and, unbelievably, when they left the owners hadn't changed the locks. I called the locksmith out the next day.

I had still only sold a handful of globes (all at under cost) and yet I now had a lease on a warehouse until 2025. That was when the Aston Martin had to go. To be honest, I'd had it almost a decade and it was too good not to share – the pleasure and the pain! I put it on eBay, a fire sale, and crossed my fingers. The winning bid was £58,000 – amazing, a £44,000 profit. As any fellow Aston owner will know, my bills over that period probably wiped out all the profit (if not the embarrassment of having spent that much on a car), but selling something on eBay for £58,000 felt pretty good. Not that I don't regret it every day. The lucky new Parisian owner arranged to come over and collect it, but I had time for one last journey. I was driving along Knightsbridge just outside Harrods when my phone rang. I answered (you could in those days). Meredith had given my mobile number to a caller. The man introduced himself, and I almost crashed the car: it was a former 007. He asked if I could deliver a globe to him in time for his wife's birthday – by the end of the following week. I obviously said yes and turned the car around to head back to the studio to make a start on the globe. With shipping we had around three days to complete the

making and painting. I had started fixing the gores to a globe the previous day so I continued with that one – I had to finish it that day to give us any chance of delivering on time. Making globes at speed had not been part of my learning process, however. The process is long, slow and tedious – necessarily and deliberately so – and there are no workarounds. I finished the gores in the early hours the following morning. Too late for any painting.

Meredith began to paint the globe the next morning, but it was quite a dull day in late September 2011. She knew the timings were tight and painted until after 10 p.m., well after dark. That was a mistake. When we arrived at the studio early the next day, the globe looked like a patchwork quilt. It's best to paint in watercolour in daytime because the pigments hide their true colours in low light. She spent two days correcting all the colour mismatches. Pavel meanwhile was making the base, and I was preparing and polishing the meridian. We managed to complete the globe and ship it all in the space of that same week.

MATCHING THE GORES

Matching the paint shades on gore panels is a supreme skill. Today we spend hours painstakingly adding layers of watercolour to shade the mountains and oceans on each individual gore. If one panel is a shade too light, you must first wet it before adding another touch of pigment. The newly wetted gore then appears darker than the one you are matching against; nevertheless, you add another layer of paint – making it seem even darker. And if you over-darken that one panel, you then have eleven other panels that are now too light and need darkening to match. Before long you are tearing your hair out. Not only that, but as we made more globes it became obvious that sometimes a seemingly perfectly matched panel would later un-match itself from adjacent panels unless we used the exact combination of pigments used originally.

We now work with the chemists at our watercolour suppliers to ensure none of our pigments fade unnaturally quickly. We also keep a book that contains every one of our in-house colour pigments and all our bespoke colours, so that we can match each one exactly. In addition, we make more than we need, so that when accidents happen we are prepared. A simple globe colour scheme will have around thirty different shades of pigment.

We'd been in Bouverie Mews for about a year when I took on an apprentice full-time globemaker, Jon Wright, who had just left college, as I needed help with the large globes. To add to the pressure, the Royal Geographical Society had just invited us to host an exhibition, its first ever on globemaking, in October 2012. We had our work cut out.

TWILIGHT

Without the atmosphere and the diffused light that it allows to reach the Earth, only objects in direct light from the sun would be illuminated, with anything else in complete darkness. So, twilight wouldn't happen, sunrise and sunset would be more immediate, and the sky would be black, as it is seen on the moon.

9

SPINNING
THE
GLOBE

BUILDING THE BASE

There is something quite magical in spinning a globe; it's impossible not to get drawn in. Perhaps it's the god-like power that you seemingly acquire; maybe it momentarily allows you to imagine that you could transport yourself around the universe; or perhaps it's simply that it allows you to explore our planet and dream of possible future journeys.

I've forgotten what it felt like when I first spun a globe as a child, but the truth is, it still gives me a thrill. And happily, it's a crucial part of the process. When we are testing our globes, we spin them rigorously to check the engineering. I also get a chance to spin our globes at speed when I am demonstrating them, but perhaps that's just to show them off.

Without exception, all visitors to the studio treat every globe like a delicate object, first requesting permission to touch it before asking, 'May I spin it?' The look of delight on their faces when they do so, very gently, is one of the greatest perks of my job. At that point, it often feels that a connection is formed, one that I am not part of. Perhaps that connection is feeling a duty of care towards something that cannot protect itself, and is looking to us to do so.

From the outset, I wanted to ensure that the interaction between the globe and its user would be an experience to cherish. Constructing the stands was therefore an important factor in the evolution of the business. This was my next challenge in globemaking: designing and building the circular tabletop, the horizontal part of the traditional wooden stand in which the finished globes would sit. This piece is not unlike the rings around Saturn at its equator, surrounding and protecting the globe. I knew from the outset that I was not the person for the complex and time-consuming task of building the stands. I'm not too bad at woodwork – I can do the basics – but I was already trying to master quite a few complex skills, and there was still a long

'I THOUGHT THIS WOULD BE EASY.
TURNING WOOD IS CERTAINLY
STRAIGHTFORWARD ENOUGH,
BUT I SOON DISCOVERED THAT THE
DESIGN ELEMENT ISN'T'

ATMOSPHERE

The atmosphere extends up to around 6,200 miles (10,000 km) above the
Earth's surface and consists of 78% nitrogen, 21% oxygen, 0.9% argon and
0.1% other gases (trace amounts of carbon dioxide, methane, water vapour and
neon). The densest part forms a layer extending up to around 5–9 miles (8–15
km) from the Earth. This is the Troposphere where most of the Earth's weather
is formed and constitutes around 80% of the mass of the atmosphere. Above
this is the Stratosphere, which extends from 12 to 31 miles (19–50 km); this is
where aircraft do most of their flying. Beyond there is the Mesosphere, then
the Thermosphere (beginning between 50–80 miles to around 340 miles),
which is where the International Space Station and satellites hang out. The
space station sits at between 200 and 240 miles, while returning space vehicles
start to warm up at around 75 miles high. The lower part of the Thermosphere
contains the Ionosphere, named because in this zone particles are ionised
by solar radiation, and this is where the light shows known as the Aurora
Borealis and Aurora Australis occur. Beyond the Thermosphere, we reach the
Exosphere (more satellites). Beyond that, you've entered outer space.

road ahead with those, so I didn't want to add another task – especially one with obvious dangers to my now indispensable fingers! There would always be many thousands of better carpenters than me even if I were to devote my life to it, and they already had the skills – now.

When Pavel was working on my house, I had instantly connected to him. He seemed the perfect person to help me construct my first tables. I talked through the design with him, then over several weeks he worked out a method and constructed a couple of the circular tabletops.

Turning the legs for the stand, on the other hand, is something I could do and is a lot of fun; I remember turning wood when I was ten or eleven at school. So I invested in an old-school lathe for the studio, and while Pavel worked on the tops, I would chisel away at bits of wood. The aim was to produce perfect legs and horizontal support beams to support the meridian.

I thought this would be easy. Turning wood is certainly straightforward enough, but I soon discovered that the design element isn't.

In Stoke Newington I had a great local timber merchant and machinery supplier called General Woodwork. The owners, two brothers and their ninety-year-old mother – who had set up the business with her husband just after the Second World War – were from a different generation and had an older mindset, and were helpful and very interested in my project. Because I was prototyping parts for the wooden stand bases, I would be down at General Woodwork purchasing items and asking the brothers for advice what felt like every other day. As I got to know them, they introduced me to their wood store, a tall, narrow building in a quiet lane behind the shop. It was enchanting, and I quickly discovered that the wood at the bottom of the piles of stock had been there decades. All the pieces were dated with the year of delivery, and I spotted some pieces

of Japanese oak that had been sitting there from just after the end of the Second World War. I spent hours pulling everything out to get to these aged pieces, agreed a price, then strapped them to the top of the car and headed back to the studio. It was a sad day when the brothers finally shut up shop in 2012. No one else in the family wanted to continue the operation, so it was replaced by yet another soulless property development. They had provided me with the perfect riposte when my new woodturners (after I suggested supplying my own timber) inevitably asked, 'Is this wood dry?'

'I think so, Harry,' I replied. 'Is sixty years of dry storage long enough?'

I was born but rarely cause any bother other than needing the odd greasing. They also have the machines to make all the patterns in wood that a spindle or column maker requires. There has been no concession to modern machinery and why should there be?

It's a close call as to who is older, Geoff and Harry or the machines, most of which belong in a museum rather than earning a crust. Until that day arrived, they agreed to make me limited runs of legs and cross-beams for my stands. Now I had the three elements – legs, tabletop and cross support – which Pavel carefully fixed together. The first rudimentary prototype base for the globe was complete.

'THE WOOD AT THE BOTTOM OF THE PILES OF STOCK HAD BEEN THERE DECADES'

Nichols Bros (Wood Turners) in Walthamstow, east London were also from a previous epoch. The firm had been founded in 1949 by the owner's father, and is now run by Geoff, along with Harry (retired 2022), his best friend since kindergarten. The premises are a rabbit warren consisting of several old lorry containers full of seven decades of patterns for spindles, a number of beautiful (and a little terrifying) open canvas- and leather-belt-driven lathes with wooden gears that run to their own beat, and a plethora of associated planers, thicknessers and saws, which have been going strong since before

Getting the meridian to fit correctly in the tabletop was another puzzle. Wood is prone to expand and contract with the seasons and will probably contract in general over time, so I had to allow enough room in the vertical notches cut from the tabletop for this to happen. I then sat the meridian within a carved horizontal notch on the support beams. I originally tried to avoid the wood-movement issue by buying old gate-leg tables for thirty pounds on eBay, but there were too many hidden nails, so after General Woodwork closed, I had to locate another good hardwood supplier.

THE SPINNING EARTH

When I spun the first completed globe in its base, a further crucial design element became obvious. The globe needs to be balanced. I won't say exactly how we do it (this is one secret you can work out for yourself) but it works in the same way as balancing a car wheel. And we can only aim for 100 per cent, friction becoming both a barrier and a tolerance, but it means that the globe naturally comes to a gentle rest after spinning. After working out the heavy point on the globe, I had to figure out how to add counterbalances – in our case lead weights – to the inside of the

Customs and Border Protection (USCBP) personnel at Kentucky airport. Perhaps they were intoxicated by its beauty or the craftsmanship (notwithstanding that I had painted this particular one), but rather than using any number of alternative methods that any normal sentient being might employ to verify the contents of an object – a phone call, a Geiger counter, an X-ray exam – they went for brute force and took a hammer and chisel to the globe.

They not only smashed it apart at the equator but somehow also warped the solid

'THEY WENT FOR BRUTE FORCE AND TOOK A HAMMER AND CHISEL TO THE GLOBE'

sphere. Initially I did this by drilling a hole from the outside and gluing in small pieces of lead before refilling the hole. I'd already discovered that it was not much fun using power tools on plaster of Paris, and no doubt there are many ways to string this particular bow, but in the end it seemed obvious to reorder the processes and add the lead prior to sticking the two halves of the sphere together.

This method of balancing the globe, or at least the material we had used to do so, created another minor issue. This I encountered in 2011 when we shipped our very first globe to the United States. It evidently attracted the attention of US

brass meridian. No mean feat. No doubt the Inspector Clouseau at USCBP thought all their Christmases had come at once and they had busted a heavy-metal smuggling operation, earning them a name check and a raise. Either way, after their excitement had died down, they very kindly did the best they could and put it all back together almost as they had found it and sent it on its way to our expectant customer. I'd like to say that they hoped nobody would notice the damage they'd caused, despite the north of South America now lining up perfectly with southern Africa, but clearly they didn't care. There was zero accountability, no explanation, no chance to even talk to the genius who did this. I wish I

LONDON PLANE

We offer many of our globe bases in locally sourced London plane. Each piece of wood is hand selected and often has a special story.

The seventeenth-century horticulturist John Tradescant the Younger 'discovered' the London plane tree in his nursery garden in Vauxhall. He believed the new species to be a hybrid of the American sycamore and the Oriental plane, both of which grew in his garden. In the late eighteenth century, at the height of the Industrial Revolution, the hardy London plane was widely planted along the city's streets as it could withstand the increasingly polluted air, heavy with soot and smoke from the new industries. Some of these original specimens are still alive today, the oldest of which, presumed to have been planted in 1789, stand in central London's Berkeley Square. Due to the trees' ability to thrive in city environments, the London plane subsequently found a home in New York parks and streets, as well as in Sydney, Melbourne, Adelaide and Perth.

The wood can be used for making plywood, veneer, flooring, furniture, carvings, chopping boards and other objects, but because the tree is not commonly found naturally in woodlands, there is a limited supply. Following the devastating storm of 1989, however, the UK had a glut of London plane wood, and then a further oversupply following the clearance of Hackney Marshes for the 2012 Olympics. The wood we're currently using comes from the clearance of the trees around Euston station to enable a new station to be built to house the HS2 rail link. We bought several of these trees in their entirety, allowing us to have the wood dried to the correct size, thus reducing waste as well as providing larger pieces of wood, meaning we can make our bigger rotational bases from a single block. Amazingly, some of this wood contains pieces of shrapnel from the Second World War, embedded in the tree trunks during the Blitz.

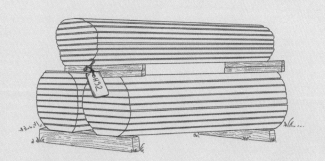

had kept a picture of the globe as it was when it arrived at the customer's house, but he was amazingly understanding and I duly made and sent a replacement.

I now insure all our shipments, and we also follow our globes on their journeys, using GPS or Bluetooth trackers. One shipper recently lost a 50-centimetre globe, packed in a five- by-three-foot crate and with handwriting on the outside detailing the sender (us) with our contact details, and the name, address and telephone number of the consignee. They lost it for a whole month, despite us sending them its precise location via our Bluetooth tracker – the crate was in a car park next to a freeway in rural Colorado. Before going for the nuclear option and buying a flight, I too turned Clouseau. Locating the address on Google Earth, I could see that it was a small retail park and there was a gymnasium closest to the location of the ping. I called the gym.

The wonderful lady on reception assured me there were no random crates in their car park, but there was a small courier company in the lot behind. She very kindly wandered over and got the company name, and I looked up their phone number.

A man called Jim picked up the phone, and I described the lost crate containing the globe. He said that, yes, the crate had been there a while (he recognised my description of the handwriting on the side), but they had not been able to deliver it as the customer's address was down a steep road. They had informed the shippers who had subbed the job out to them and who had found it easier to call it lost because it was my insurance that would pay. I then put the customer in contact with the courier and they arranged delivery between themselves. A week or two later the original shipping company emailed to say they had found the lost crate and later even had the audacity to send a bill.

GLACIERS

The Earth has been much hotter and much colder over its history. At times the entire planet has been frozen. The last full ice age ended around 11,000 years ago, with its maximum extent around 20,000 years ago. We are now in an interglacial period.

During the mini-ice age of the fifteenth and sixteenth centuries, glaciers covered around 33% of the Earth's surface, compared to around 10% today. In the UK from 1608 until 1814, a frost fair was held in London almost annually on the River Thames, as the river would freeze for up to two months. In 1683–4, the winter temperature was so severe that the sea off the coast of southern Britain froze up to two miles offshore.

Alaska has an estimated 100,000 glaciers, while 80% of Greenland is covered by a single glacier or ice sheet. It measures 1,710,000 sq km (660,000 sq miles), second only to the Antarctic sheet, which was formed at least 40 million years ago.

SUPER-CONTINENTS AND TECTONIC PLATES

Pangaea is the name given to one of the most recent of the super-continents that have formed on Earth, and consisted of almost all current land masses. It was the first to be credibly reconstructed by geologists. Many current continents fit neatly into each other's coastlines and share identical rock formations where they previously met. The theory of plate tectonics governs the movement of the world's land masses. The other two more recent super-continents have been named Gondwana and Pannotia. There is general agreement amongst geologists that there were probably three before these, which are named Nuna, Rodinia and Ur.

We are currently in an unfriendly period where the continents on the whole are all drifting apart, but since the Earth is spherical, it is only a matter of time before they have enforced intimacy again. Come back in several hundred million years.

The Earth's crust is constructed of a series of (tectonic) plates which either move apart, like the North Atlantic Ridge, creating islands, archipelagos and new land masses, or move towards each other. This can have several consequences: creating huge mountain ranges like the Himalayas as both plates refuse to yield, or a range on one side and a huge trench (subduction zone) on the other, like the Andes and Atacama Trench. When they slide alongside each other, like the San Andreas fault in California, earthquakes and volcanoes often result from these movements. These earthquakes are frequently very minor and allow pressures to equalise over many years and centuries, but inevitably these pressures build up, and when they are released, the Earth's crust can move by many tens of metres. If this happens underwater it can create huge tidal waves (tsunamis), and if this happens on land, it can cause extensive damage to man-made infrastructure, such as in Japan in 2011 and Turkey/Syria in 2023.

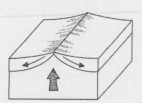  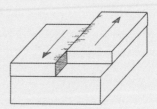

10

ORGANIC GROWTH

The one thing that has been consistent from the beginning is how we have grown our product range. When I conceived my very first 50-centimetre globe, I hadn't even decided on the design for the stand, more especially the profile of the table leg; I just put a chunk of wood in the lathe and twenty minutes later I had my pattern. I had obviously looked at examples, but a quick search on Google will show you that globemakers over the years have evidently struggled with their stands – even the masterly Blaeu, whose stands, like ours, vary. An online search will reveal many poor designs, which can be quite challenging on the eye.

As we became more established, our client base grew and came to include designers and architects who have contributed their own ideas. It both challenges us and makes the job even more interesting to take on a project that requires an original design, and new engineering and construction elements, and we often have to make fresh jigs.

'CUSTOMERS WILL GO TO GREAT LENGTHS TO ENSURE THEIR PURCHASE REACHES THE CORRECT SPOT'

Both our 80-centimetre and 127-centimetre globes were the result of my inability to say no to a new and potentially exciting but at the same time daunting or even apparently impossible project, and customers often ask if we can make a globe just a little bit bigger or smaller to fit in their house. When a new client places an order with us, we will stress the size of a globe and remind them to measure entry points and where they plan to place the globe.

A globe, particularly a large one, may well be the most important feature in a room, and we have discovered that customers will go to great lengths to ensure their purchase reaches the correct spot. One Spanish customer had a wall demolished in his medieval castle in preparation for the globe's arrival, so that he could get it into his library. The wall was then rebuilt, and the globe is now stuck there until the same process is repeated. We have had to cut the feet off a base when one customer informed us that he had decided on a different location for his globe than originally planned. In one extreme example, one of our Churchill globes was winched twenty storeys up the outside of a skyscraper to reach its designated location.

THE EQUINOXES

Twice a year, in March and in September, the sun is directly above the equator, and hence every point on the Earth records twelve hours of sunlight and twelve hours of darkness.

The Earth's rotation axis remains tilted at the same angle as the Earth orbits the sun, so after the March equinox (the Vernal Equinox), the Earth begins to move to a position in its orbit where this tilt brings the Northern Hemisphere closer to the sun. This marks the beginning of spring. In the Southern Hemisphere, meanwhile, the opposite occurs as the Earth tilts away from the Sun: this is the Autumnal Equinox, marking the beginning of autumn. In September, the Vernal Equinox occurs in the Southern Hemisphere, while in the Northern Hemisphere, inhabitants experience the Autumnal Equinox, the days grow shorter as winter draws near.

The arrival of the Vernal Equinox has long been a reason for celebration across many civilisations. The sphinx at Giza in Egypt was built to face directly towards the point at which the Sun rises on that date. In Christianity, Easter falls on the first Sunday after the first full moon after the Vernal Equinox, which accounts for the irregularity of its date.

HOW TO GET INTO HARRODS

The first globes I made in 2010, were, at 50 centimetres in diameter, too big to offer to most shops. In fact, I had never really thought about retail outlets, but then some time in 2011, Frank, a Venezuelan architect and photographer, dropped into the shop. He stayed for hours, then came in again the next day and the day after that. During our chats, Frank suggested that I should start to make smaller globes. It wasn't a radical thought. Most people expressing interest in globes enquired about desk models after all, and at this stage I was marketing only the 50-centimetre and 127-centimetre globes, so why not introduce a new desktop

about learning the process. I wanted the 22-centimetre model to spin on bearings, and for prototyping I happened to have a spare fibreglass 22-centimetre sphere in the studio. I turned a little base and ordered some miniature bearings, which I fitted into it. The sphere, though, was very light, so it rotated awkwardly on its bearings and flew off if spun too quickly.

Frank and I talked about options, and he brought up my previous career. I had been regaling him with stories of famous people I had ejected from the bowling club, and he came up with an idea. Why not get some bowling balls and fill in the finger

'IT ROTATED AWKWARDLY ON ITS BEARINGS AND FLEW OFF IF SPUN TOO QUICKLY'

model? But at that point in the evolution of Bellerby & Co., I was in almost as much trouble as the central banks. I took a leaf out of their book however and threw caution to the wind. Despite only having sold a handful of 50-centimetre models at a substantial loss, I had the Yinka Shonibare studio and Churchill commissions in my pocket, so we started development of a new 22-centimetre (8.7-inch) globe.

Frank and I made a deal: I would pay him a basic salary with the promise of a good bonus per finished globe, and he set

holes? Frank, you're a genius. Why indeed not? I got in touch with the manufacturer and commissioned them to make bespoke spheres to our specification, in plain black or white. Even though our spheres were a little larger, I reminded them that they should forego the final procedure of drilling finger holes. I just needed perfect spheres and they after all were eminently suited for the task. While the minimum order was several hundred I had advance orders for forty and was confident I could sell a good few more.

Later that year I was reading an article in the London *Evening Standard* about an art event that a charity called the Elephant Family was organising. They were asking artists and designers to decorate large eggs for an Easter egg hunt across London the following year. I thought I would give it a go and create an 'Egg Globe'. By the time I took delivery of the giant egg, however, I had just under a week to apply and paint the bespoke map. While the southern hemisphere was straightforward, for the northern hemisphere I needed to warp each gore individually using Photoshop to make it fit the ovoid shape. It was worth it – I was very proud of my finished egg-shaped globe, which the Elephant Family later sold in a live online auction. A further model complete with a one-off brass meridian was made for the charity's New York Big Egg Hunt in 2014, and was sold at a star-studded auction at Christie's.

In early summer 2013, however, we had sold just one globe in two months, and I remember thinking, this isn't going anywhere. If it doesn't get much better,

I'm going to have to work out a new way of making money and just keep this as a hobby. But the lull turned out to be just that, because it was then that I decided that we should try to sell our globes in Harrods. This was not straightforward. For a start, which department would sell globes? Books, maps, luxury goods? And how do you begin to hunt down the appropriate buyer? At a loss, I found the name of Harrods' chief executive. I packaged up a globe, addressed it to her, and one evening dropped it off at the store's goods-in loading bay.

I had no idea if I would get a response, if I did what it would be, or even if the globe would reach its addressee, so I was amazed when at 6 a.m. the next morning, I received an email from her almost ordering me to come in to discuss a first order. The only problem was that the unit price I had suggested didn't give us any profit margin, but that was something I could work on long term. I figured that Harrods agreeing to stock our product was still a major coup for any new company, and so I put it down to advertising spend.

In 2013, I finally persuaded Jade that she should leave her job. She worked for a music venue whose owner was a less than ideal employer. It was only a matter of time before she left, though she did manage five long years and learned a lot.

After a couple of months at home recharging her batteries, she suggested she come into the workshop and help – perhaps answering the phones, she suggested. I agreed but didn't let on that this would likely be no more than a couple of times a week, figuring that she would quickly find other things to do, maybe even learn to make globes. However, after staring for some time at the silent phone, she took it upon herself to look at our social media and general online presence. I had a Facebook page, an Instagram account and a functioning website.

Instagram had launched in October 2010, so it was a relatively new platform and was then a very positive one-way user experience with no adverts, allowing the user to scroll through in spare moments for inspiration. Jade started taking photos of the studio and posting them once or twice a day, as well as finding photos from shoots that I had done in the past few years but then forgotten about. To begin with she gained a few followers here and there, but I didn't realise she had a longer-term plan. She would spend hours hunting down the Instagram accounts of editors, feature writers and influencers (a new term back then), then follow and like and comment on their posts. Instagram was a great source of content for magazines, and quite quickly we were deluged with requests for pictures of the studio, and me. She also reached out to the Instagram team itself.

At the same time she found a video that I had shot with two young filmmakers, Arran Busk and Jamie Smith, in 2011/12, which had been sitting on Vimeo, the video content website, with no more than a couple of hundred views. She contacted Vimeo's content editors to see if they would

THE SPINNING EARTH

The Earth's rotation axis is not fixed, but has been drifting towards North America for much of the twentieth century. The drift is, however, measured in centimetres per year so has no noticeable effect, but needs to be taken into account for GPS measurements. It is mainly caused by the movement of water around the planet but also by continuing movement in the Earth's mantle caused by the last ice age. In the year 2000, the movement began to veer in direction towards the United Kingdom at a rate of around 7 inches (18 cm) per year. This change was mainly caused by a loss of ice in Greenland and Antarctica, but also as a result of drought and a subsequent loss of water in aquifers.

feature the film on their site as a 'staff pick'. At the end of August 2013 they did. They also tweeted the link to their 122,000 followers. Within weeks, views of our little film exploded from an original 250 or so to over 200,000.

I hadn't spoken to Arran for a couple of years. He called me out of the blue.

'Pete, what's going on with the film?'

'I've no idea,' I replied. I genuinely didn't. Jade didn't keep me posted on what she was doing.

'It's been chosen as a staff pick on Vimeo, and it's getting tens of thousands of views.'

'Is that a good thing?' I asked, slightly concerned that he might have an issue with this sudden development and demand some compensation!

One cold January, we had spent three days filming the video (he sure wasn't getting paid for putting me through that freezing hell!) and neither he nor I had done anything with it since. In the space of a few weeks Jade had gained it more exposure than either one of us had done in several years or indeed was capable of.

Instagram, meanwhile, following Jade's intervention, also highlighted Bellerby & Co., #globemakers, as an account to follow on their app. In the space of a couple of hours while we were enjoying a Sunday pub roast with some friends – who were a little perplexed at our excitement – our followers shot up to just shy of 100,000. Both directly and indirectly, Instagram showcased us to a whole new audience that we would otherwise not have found, and we were inundated with requests for interviews from magazines and TV, so even though I hate being filmed or photographed, I was now left with little choice. Jade, at the age of twelve and five feet ten inches tall, had been an even more reluctant model, though she had been paid quite handsomely for her time, and I think she was secretly amused by my discomfort. She knew, however, that I would rather be making globes. From around this time, I was photographed at least once a week, sometimes two or three times, for more than two years. We shot any number of videos for TV companies from all over the world, which I discovered would take between one and three days' filming for just a few minutes of broadcast footage. It was relentless.

I want to. This led to continuity problems, but if I just repeated the same lines time and again, it would feel like acting and my delivery would likely turn to cardboard. After washing and ironing a shirt three times for one job to maintain continuity for one filming job, I decided to adopt the Steve Jobs approach to clothing. That meant fifteen identical blue shirts.

Thanks to Instagram and to the ensuing publicity all over the world our sales began to grow rapidly. Our timing was right in that artisan crafts in general had been gaining in popularity around the world. After years of buying cheap throwaway items, increasing concerns about sustainability meant consumers were becoming keener to invest in something that was unique, well made by committed craftspeople and with a special meaning for them. Ultimately, and

'A GLOBE IS A CONSTANT REMINDER OF WHERE THEY ARE, WHERE THEY WANT TO BE AND WHAT THEY HOPE TO ACHIEVE'

As many of the filmmakers who interviewed me would probably attest, I can be a challenge to interview. When asked a question about the company, I would give an appropriate answer. But if the light changed or a plane flew overhead and I had to do the take again, I would give a different answer – not factually different, just differently phrased. I planned nothing, least of all the interviews. I wasn't spending time beforehand rehearsing; I was just giving truthful answers. But given that answers on globemaking can be long, I was neither able to memorise what I'd been saying, nor did

unintentionally, I think we had helped to make globes relevant again. And just like me, a lot of people like to daydream about exploring the world, and a globe is a constant reminder of where they are, where they want to be and what they hope to achieve in life. We also started to get requests from production companies to use our studio for filming or to rent globes for productions. My favourite was when Morgan Freeman and Mohamed ElBaradei visited to film for Freeman's series *The Story of God*. We kept it quiet, but following a catered lunch at a local restaurant we suddenly had a crowd and the

game was up. Filmmakers and their teams really are the most competent, organised and unflappable people. If you want something doing and need people who can cope with a crisis, hire a film crew.

As our orders increased, Jade had another idea. 'Why don't we personalise our globes?' she said. Instead of every one having identical cartography, we could add bespoke individual details, be it small additions – just a few smaller towns highlighted – or large, with hundreds of illustrations, travel routes and edits. I wasn't keen and had my reasons. I was at this stage still in charge of sales and had it all organised perfectly. The right side of my brain kept all the creative details (at this stage, the colour of the globe and the chosen wood for the base). And the left side of my brain held the account details, customer names, etc. Under pressure to explain my reluctance to complicate our output and increase demands on my limited

time, I explained that the lack of organised paperwork was my reason.

Jade offered to take over responsibility for sales to help spread the workload. I agreed. It was only when she asked me to do the handover that she fully realised why I'd been so reluctant about coming clean – it was nearly all filed in my head. The feedback from customers after Jade took on the sales role was immediate and very positive.

By 2018 we were turning about 350–400 globes per year, and the waiting list was between six months and two years depending on globe size. Since then we have grown gradually, slowly expanding but wary of overstretching ourselves. We gained a Queen's Award for Export in 2018 and 2021, and business continued to increase steadily. Without Jade's intervention and imagination, the company would be a dramatically smaller version of what it has become.

WHY WE CAN'T MAKE TOO MANY GLOBES IN ADVANCE

Borders change, countries change. In 2008, for example, Sudan split into Sudan and South Sudan, along with eventually deciding on a new capital for South Sudan (Juba), so we had some globes where Juba was in parentheses. Swaziland became Eswatini; Republic of Cabo Verde formerly Republic of Cape Verde; Macedonia became North Macedonia; Turkey is now Türkiye.

In addition, regions and states change, even mountain names change.

It's not that often, but it immediately dates a globe and would a customer want to buy something out of date? Though on the other hand sometimes that's exactly what someone wants.

HOW TO DATE A GLOBE

When I started this project I had not really thought about this question.
It is surprisingly easy. Over the centuries there have been many changes to
cartography, from country names and borders, especially relating to colonisation
and its subsequent reverse, to changes in capital and city names, region and state
names, and physical geography changes. For example, in the 1930s alone the
following happened:

1930: Constantinople, long known colloquially by the name, officially becomes
Istanbul

1931: Japan invades Manchuria and renames it Manchukuo at the end of the war
in 1932

1932: Saudi Arabia founded

1932: Kingdown of Iraq becomes independent from Britain, becoming known
officially as the Hashemite Kingdom of Iraq

1933: Nazis establish totalitarian control in Germany, and rename the country
the German Reich

1934: Italian East Africa merges Somaliland, Eritrea and Ethiopia

1935: Persia becomes known officially as Iran

1938: Germany annexes Austria into the German Reich

1938: Bolivia and Paraguay sign a treaty agreeing to the split of the Gran Chaco region

1939: The Sudetenland in Czechoslovakia annexed by Nazi Germany; in 1939
the state is broken up and Bohemia and Moravia become a protectorate
of the German Reich, and the Slovak Republic becomes a client state

1939: Germany invades Poland

1939: Hatay State, formely part of the French Mandate for Syria and the Lebanon,
annexed by Turkey after a referendum

It may have been a decade with many changes, but for most of the twentieth
century, you can date a globe to within a couple of years of manufacture. The
proviso is that we are talking here about the date that a globe was designed and
printed. Sometimes globemakers may print many years' worth of stock to make
the project viable, so you are dating the time the map was created and printed
rather than when the physical product was manufactured. A single sheet of gores
takes up considerably less room than a globe, so they are much easier to store.
See also 'Why We Can't Make Too Many Globes in Advance' on page 196.

KARMA

Then came 2020 and Covid-19. As the virus wreaked havoc, and much of normal life and industry ground to a halt, I imagined that there would be little temptation for people to spend on items such as bespoke globes. I firmly expected our sales to fall off a cliff. In March and early April 2020, they did.

Lockdown was a tough time for everyone, but as a small business we were extremely fortunate. We had a six-month order book and a huge, well-ventilated (draughty!) studio where up to four people could continue to work on a percentage of our large globes while keeping a more-than-safe social distance. Over half the team were on furlough; others worked from home making and painting smaller globes

cannot praise every one of them highly enough.

The company has since recovered well, though with wild variations between territories in the order book that are just impossible to understand. So, it can still sometimes feel like we are not quite a sustainable business. We don't advertise, so live by word of mouth, and ensuring that we are well enough known that we become self-perpetuating is the goal. Maybe we are there, who knows?

We have, of course, come up against some competitors along the way. A couple of operations have started up in Europe, one trying to mimic our paint colours almost identically. We even tracked down a business in China that was producing a mass-market imitation of our

'IT'S IN DIFFICULT TIMES THAT A PERSON'S TRUE COLOURS EMERGE'

in their kitchens and studios, with people gradually returning over the course of the pandemic. I meanwhile became the company driver, my responsibility being to transport globes between team members' houses and HQ in Stoke Newington on London's empty roads. While I concentrated on this task, I also took the opportunity to delegate. This allowed the company to mature rapidly from a model in which I was running everything to having managers running everything.

It was undoubtedly a stressful and heartbreaking time for the world in general, and I am just grateful that my team and their families stayed healthy and were able to maintain their usual positive can-do mindset. It's in difficult times that a person's true colours emerge, and I

mini desk globe using poorly chosen materials and unreliable bearings (we purchased one for fun). We read their elaborate story suggesting that they had set up their studio decades earlier, but struggled to find any reliable evidence.

Then there are commercial makers around the world – in the US, Germany, Italy (still making those wonderfully kitsch 1970s bar globes), Japan and a few other countries including Russia and North Korea. But I think most consumers can tell when there is no passion behind a product. And nothing can compare with a globe made by hand, with love and care. There are not that many markets with gaps in them, I suppose, but even in a saturated field there is always room to put your personal stamp on what you are creating.

EPILOGUE

Based in a studio with twenty-five talented and dedicated people working to create wonderful globes that ship worldwide, Bellerby & Co. Globemakers is beginning to feel like an established company, rather than the rudderless, madcap start-up operating from my dining room it once was. Many studio visitors assume that my father passed it down to me. We now make over 600 globes of varying sizes and types each year. That Vimeo video has been viewed over 750,000 times and we now have more than 150,000 followers on Instagram.

We continue to strive for perfection, keep abreast of cartographical and political changes and are constantly updating our maps. For each new globe, we print a new, up-to-the-minute map to order, with the year of completion on the cartouche, so it is as accurate as possible a representation of the world at the time of making. Every day we also update our list of world leaders, which on request we print on the map in the middle of the Pacific Ocean. Our printing set-up gives us enormous scope for customisation and to add personal touches – for example, we can add the name of a home town, even a tiny village, on request. We are also regularly asked to feature customers' artwork and illustrations of sentimental significance, some more idiosyncratic than others: travel routes, portraits, deckchairs, pets. One Hong Kong customer commissioned us to illustrate four of his dogs piloting biplanes, scarves streaming behind them, and a fifth floating nearby under a parachute. Another client, who was flying to Antarctica, insisted we include the airstrip where he was planning to land his plane. One of our new recruits after several months went as far as to describe her work as a painter as a 'unicorn job'.

TIME

Over the centuries the Earth, or more specifically the length of its rotation and the simplicity of its measurement, has been used as a very precise measure of time. The Earth rotates once every 24 hours with respect to the sun, and once every 23 hours, 56 minutes and 4.091 seconds when measured against the stars. This is known as a sidereal day (see p44), which is a timekeeping system used by astronomers. While there are now more accurate measurements of time (measured in seconds) involving oscillations of caesium atoms, the Earth's rotation and the sun's position in the sky have always been a simple and reliable method of timekeeping.

Our celestial globes have also taken on a life of their own, which brings a whole new challenge – how to keep the map of the stars accurate. Stars move by a tiny amount in the sky, so celestial globes are dated in epochs, though it takes years to notice a difference. Nevertheless, we use a contact at the European Space Agency for the most up-to-date information.

It has been quite an adventure. At the age of forty-three, when I should have been deeply involved in a 'proper' career and had a family, I fell into a unique profession. I have been incredibly lucky along the way. Many things have unexpectedly contributed to the success of the business, and perhaps you do make your own luck. It certainly feels at times as if karma is on my side, but perhaps my bloody-mindedness has also played a part. I love solving problems with a positive mindset, and this has perhaps been my most useful asset throughout. Having said that, I could not have done this at a younger age. What I learned in other jobs or hobbies over the years has been crucial in helping this project succeed.

While some things, like the advancement of printing and digital mapping, have helped the growth of the company, it's a sad fact that Harry, the turner, has retired and both Manor Foundry and the two local wood stores have closed down since I started writing this book.

There will no doubt be more challenges in the company's future, but it has been anything but boring. Just being in the creative workspace of the Bellerby & Co. studio with a team of brilliantly talented artists and craftspeople is reward enough, but creating beautiful globes and knowing the pleasure they bring to people is amazing. There is never a morning when I dread going to work, and I can't imagine that day ever arriving. It's certainly quiet work – sometimes hours can go by in complete silence as we get lost in the rhythm of the movements and detail – but it's never solitary. As we work in teams in small areas, there is still a lot of interaction.

but also a feeling of protectiveness towards the planet.

And creating globes brings with it a constant reminder of the greatest challenge facing humanity, one on which our leaders will for ever be judged. As we go carefully about our work, the fragility of the globes reminds me daily of the uniqueness of our own existence, the wonder and beauty of our planet, the pressing need to protect it from further desolation. On a personal level, it also drives home to me the need to take advantage of every day that we are here on this Earth.

My passion for globes and globemaking remains undiminished. Finishing any globe, little or large, but perhaps especially finally completing a Churchill, is always a moment of pride. It is also bittersweet because it can be quite a wrench to pack it up and send it out into the big wide world where it might come to harm. It can often feel like saying goodbye to one of your own children.

'THERE IS NEVER A MORNING WHERE I DREAD GOING TO WORK, AND I CAN'T IMAGINE THE DAY EVER ARRIVING'

Walking into the studio every day, and seeing anything between ten and twenty globes in various stages of creation, is always an uplifting moment. On the one hand, you have a sense of the individual customer who awaits each globe and the progress that is being made; on the other, you experience a similar feeling to the overview effect that astronauts recount. Seeing the Earth (albeit recreations) standing there alone in a silent studio gives you a sense of calm,

Finally, in 2010 – and after officially incorporating Bellerby & Co. Globemakers – I gave my father his first globe. It was yellow-brown, harking back to seventeenth-century globe colour schemes. 'So that's what you've been working on the last year or two,' I remember my father saying dryly. I then placed the globe in the living room beside his armchair, where he read the paper, watched television and enjoyed afternoon tea (and evening gin).

THANK YOU

Some of these people are not mentioned in the book, but they know how valuable they were to me.

FOR THE BOOK

Jade Fenster (my partner), for her fantastic work and support over the years - and for helping with the vision and organisation needed to make this book possible. Thank you, also, for understanding the company ethos from the outset and pushing me to challenge myself.

Emma Bal, for talking me in to writing.

Rose Davidson, for taking me back to school.

Dave Brown, for the beautiful layout.

All the wonderful (and patient) team at Bloomsbury.

FOR THE JOURNEY

Kelly Tregenna, who spent days with me at the beginning learning to plaster cast.

John Essex, the map morpher.

Frank Balbi, who knocked on the front door and stayed for a year.

Camilla Holder, my first painter who *also* knocked on the front door and stayed for a year.

Meredith Owen, for painting globes in the early days.

James Harrison, for the much needed early woodworking skills and help with engineering challenges.

And everyone else who had faith and invested time and effort in those early days when we had a sparse order book, and a lack of heating in the winter and air conditioners in the summer!

Thank you to all the 'globers' team past and present for helping create my vision, helping to solve the many unexpected challenges over the years, for understanding my unconventional ways, for listening to my tangents, for helping create the most amazing studio, epic summer BBQ's and keeping the dance floor going well past my bedtime! I am very lucky to be surrounded by talented artists with so many differing and unique skills who make the company feel like a family.

SPECIAL THANKS TO...

Isis Linguanotto, who started as a watercolour artist ten years ago and now also helps across so many fields as we have grown and needed so much more support! N.B. she now runs the studio.

Eddy Da Silva Fernandes, for always challenging himself, and becoming an extraordinary Head Globemaker.

Jon Wright, my first Apprentice Globemaker who came straight out of art college and went on for nine years to help solve numerous technical issues.

Leonardo Frigo, for pushing boundaries in making. Also for hiding the fact for ages that when he first came to the studio he just wanted to look around, but such was his English (none) and my Italian (none), I thought he was asking for a job; we required help in the wood-working department at the time (there wasn't one) - and so he started the next day.

Ed Burgass, for being a sounding board and helping me in the discovery of local ale houses.

Arthur De Borman, Sam Tidman, Tim Gibson and all the AD's, who rented space when we first moved to Bouverie Mews. The most out of control people I have ever met.

Kirsty (and George who became the de facto 'studio dog'), for starring in a wonderful news programme with Simon Reeve (documentarist) about apprenticeships in our company - on her very first day in the job (possibly with a lack of warning!). She, along with George, are still here eight years later.

FAMILY

Mortimer, my beautiful cat who oversaw all the early prototypes and was always on hand to help me find my kitchen.

Linda, Alan and Genevieve Fenster, for always being my biggest fans.

Mum, for cajoling Dad to say thank you.

SUBBIES

Michael at City Printing.

Charlie Briggs and his curved works of aluminium, and my numerous visits to the Aston Martin factory.

Harry and Geoff at Nichols Brothers Woodturners (Harry retired in 2022).

The team from London Woodwork (closed down).

Ruth, our engraver.

Bruce and his beautiful London Plane timber at Fallen and Felled.

S & M Timber & DIY on Green Lanes (closed down).

John and Jason at Manor Foundry (closed down).

James Mosley, who knocked on the door and lent me his personal font.

Peter Barber, who despite being head of maps and topography at the British Library, moonlighted as a wonderful interpreter for Italian TV.

OTHERS

Euan Myles, for capturing so many beautiful moments in the studio with so much positivity and helping at a moment's notice, despite living in Scotland.

Chris and Janice Kohut, for their early support and enthusiasm.

The Royal Geographical Society, for their continued support.

All the wonderful photographers who have spent time here, the TV channels and features editors - CBS Sunday Morning especially. Without them all we would still be a team of three.

The 2008 financial crisis - and the Bank of England's interest rate panel for their adept handling of the macro economy in 2008.

INDEX

PHOTO CREDITS

Alun Callender: 138; **Ana Santl for Ignant:** 12–13; **Andy Donohoe:** 134; **Andy Lockley:** XVI–XVII, 54–5, 145, 202–3; **Aron Klein:** 137; **Carmel King:** 11, 67, 167, 171, 197; **Charlotte Schreiber:** 183; **Clarisse d'Arcimoles:** 184; **Dave Brown:** 35, 36, 45, 81, 150–1, 200; **Euan Myles:** IV–V, VI–VII, X, 3, 8, 16, 21, 25, 39, 48, 52, 56, 62, 76–7, 78, 85, 100, 109, 118, 125, 146, 152–3, 159, 160–1, 168–9, 174, 177, 178–9, 188–9, 190, 209; **Gayan Benedict:** 70; **Getty Images:** 26–7, 94; **Guido Bollino:** 40, 90, 92–3, 99, 180; **Hal Shinnie:** 127, 131; **Harry Mitchell:** 86–7, 212; **Johnathan Swann:** 195; **Justin Ratcliffe:** 103–4; **Kalpesh Lathigra:** 165; **Kaori Oyama for Beams & Co.:** 220–1; **Kimisa H:** XVIII; **Nic Crilly Hargraves:** 28; **Olga Changunava:** 106; **Owen Harvey:** 4, 115; **Paul Marc Mitchell:** 112, 192, 199; **Ruth Anthony:** 122; **Sebastian Boettcher:** 204, 206; **Tanja Schimpl:** 148; **Toby Essex:** 30–1, 132–3, 210–11; **Tom Bunning:** 32, 73, 102, 116–17, 143, 155, 162, 187.

A NOTE ON THE AUTHOR

Peter Bellerby is a world-renowned globemaker and founder of artisan globemakers Bellerby & Co., the only truly bespoke makers of globes in the world. After a successful career in television, in 2008 Peter began his search for a special globe for his father's eightieth birthday. Not finding what he wanted, he set about making a globe himself by hand, which was the start of an arduous and fantastic journey that led to the founding of his company in 2010. Bellerby & Co. Globemakers have won the Queen's Award for Enterprise in the international trade category twice, in 2018 and 2021.

@globemakers | bellerbyandco.com

BLOOMSBURY PUBLISHING
Bloomsbury Publishing Plc
50 Bedford Square, London, WC1B 3DP, UK
29 Earlsfort Terrace, Dublin 2, Ireland

BLOOMSBURY, BLOOMSBURY PUBLISHING and the Diana logo are trademarks of
Bloomsbury Publishing Plc

First published in Great Britain 2023

A catalogue record for this book is available from the British Library

ISBN: HB: 978-1-5266-5087-0; eBook: 978-1-5266-5090-0; ePDF: 978-1-5266-5088-7

2 4 6 8 10 9 7 5 3 1

Commissioning Editor: Rowan Yapp
Project Editor: Faye Robinson
Design: Dave Brown, apeinc.co.uk

Printed and bound by Graphicom, Italy

To find out more about our authors and books visit www.bloomsbury.com
and sign up for our newsletters

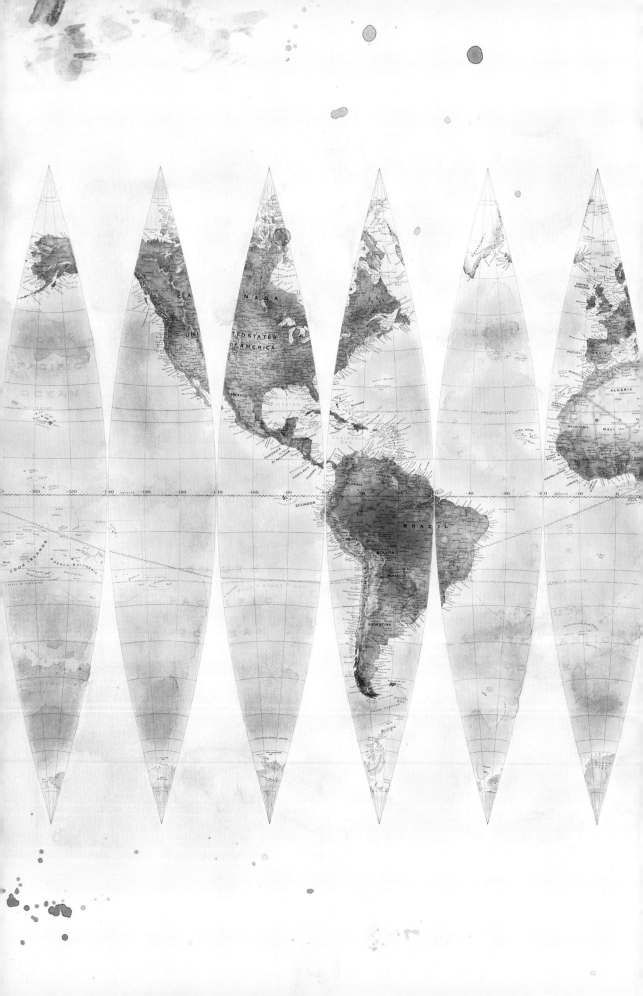